The TRILOBITE Book

THE TRILOBITE BOOK

A Visual Journey

RICCARDO LEVI-SETTI

The University of Chicago Press
Chicago and London

RICCARDO LEVI-SETTI, professor emeritus of physics at the University of Chicago, has served as director of the Enrico Fermi Institute and as an honorary research associate at the Field Museum, Chicago.

The University of Chicago Press, Chicago 60637
The University of Chicago Press, Ltd., London

Printed in China

33 32 31 30 29 28 27 26 25 24   2 3 4 5

ISBN-13: 978-0-226-12441-4 (cloth)
ISBN-13: 978-0-226-12455-1 (e-book)
DOI: 10.7208/chicago/9780226124551.001.0001

Library of Congress Cataloging-in-Publication Data

Levi-Setti, Riccardo, author.
The trilobite book : a visual journey / Riccardo Lvei-Setti.
pages cm
Includes bibliographical references and index.
ISBN 978-0-226-12441-4 (cloth : alk. paper) — ISBN 978-0-226-12455-1 (e-book) 1. Trilobites—Pictorial works. I. Title.
QE821.L455 2014
565'.39—dc23
2013028730

∞ This paper meets the requirements of ANSI/NISO Z39.48-1992 (Permanence of Paper).

CONTENTS

PREFACE

It took me approximately 15 years to put together the material included in the first edition of my book, *Trilobites, a Photographic Atlas,* and see it published by the University of Chicago Press in 1975. The task included, first of all, the fun part—the treasure hunt—digging out the specimens from their burial rock at several locations around the world. Looking for the best-preserved find was often a frustrating enterprise. Once the specimens were brought home, the unwanted cover matrix had to be chipped away, typically under the optical microscope with a variety of tools; and often the specimens had to be prepared for photography using appropriate techniques to enhance their appearance. Then came the photography, at that time using close-ups with fine-grain black-and-white film, under well-studied illumination, with my Leica cameras and, when needed, magnifying aids. It took some 20 or 30 shots to produce a view that would satisfy my taste. Developing the film and printing the photographs were also parts of my repertoire, using all the tricks of the trade. The fieldwork was enriched by loans from musty museum depositories and by borrowing samples from collector friends. In parallel with the assembly of the pictures, the paleontological identification of the material had to be sorted out of the technical literature, and many hours were spent in the library. Writing the text to accompany the selected images and editing the work were laborious, especially prior to the availability of word processors. The goal of the book was to entice a sizeable audience of amateur fossil collectors, students and professional paleontologists alike—by offering a blend of visual experience and technical significance.

The positive reviews as well as the criticisms raised by my first edition were of great help in the realization of a second edition of my book, *Trilobites*, which took another some 15 years of gestation, appearing in 1993. This involved much traveling to new and classic fossil localities and world-renowned museums, to cast a broader net of coverage, while reducing the duplications among my own findings. This led to repeated memorable visits to the Barrandian of Bohemia, to Swedish localities and museums, to Wales and Scotland, and to yearly excavations in the Manuels River gorge of Newfoundland. The casting of the net was greatly enhanced by an explosion of trilobite popularity at the Tucson Gem and Mineral Show, an annual early February event. Such explosion in interest was greatly fostered by the appearance of a host of beautifully preserved giant trilobites from Morocco, soon to become the mecca for trilobite collectors. From a technical standpoint, the second edition still had to rely on the demanding processes of black-and-white photography, much to my dismay because it was unable to convey, with the exception of the cover pictures, the excitement of color reproductions. The use of word processors was still limited to floppy disks for the written word, and all images had to be reproduced from hard copies from my darkroom. Nevertheless, my second edition approached the distinction of becoming a classic of the amateur trilobite literature, as a search on the web, 20 years after its appearance, can testify. Much happened during the years since my second edition was published. This latest period saw the advent of digital photography, Photoshop, and megastorage of digital images, including some 15 gigabytes of color trilobite pictures in my laptop and pocket Firelight.

The time had come to satisfy my long-standing wish to document at its fullest the visual experience of looking at trilobites as they are exposed from their burial, in their often stunning natural color. It had finally become feasible to publish a color edition of a trilobite book, which is not a third edition of my series but a new book called *The Trilobite Book: A Visual Journey.* My approach to this new book was closer to my experiences in the field, based (whenever possible) on my travels to a number of important fossil sites or on images from collections of consenting friends from localities beyond my reach. Thus, it was possible for me to introduce an anecdotal element to my quest for new specimens from selected hunting grounds. Among many others worth highlighting were my encounters with the precious vestiges of the discoveries of Joachim Barrande in Prague and Bohemia, introduced to me by several friendly paleontologists of the Czech Republic. To obtain an unadulterated collection of the varied and exceptionally well-preserved trilobite fauna of Morocco, I opted for a firsthand approach, by making more than 11 yearly two-week field trips to the Anti-Atlas and sub-Saharan ranges. Here, I found the first and second editions of my trilobite book to be very popular, even among illiterate Berber collectors, who guided me to impervious fossil localities and showered me with gifts. This was a welcome break from the cost-inaccessible specimens seen at the Tucson Show, even if, at times, I could only bring back indelible memories and loads of pictures. The value of the Tucson Show should not be minimized. Fossil dealers bring to Tucson, among a multitude of expertly prepared trilobites, enormous slabs from Morocco containing unprecedented assemblages of trilobites that no museum can afford to purchase. These shed light on various aspects of the trilobite's life habits and the large-scale depositional environments where trilobites lived, often providing snapshots of the richly inhabited primordial seafloors, similar to that which nowadays can only be seen by scuba divers. The Tucson Show also brings together trilobite specialists from all over the world, with their treasures, and offers me the widest casting net I could ever imagine possible—and many new friends. Among these new resources was the Saint Petersburg Paleontological Laboratory; although I never traveled to Russia, I was fortunate to be allowed by the laboratory to include a number of images derived from the exquisite professional pictures of their superbly prepared spiny trilobites. Captivating coverage of their work and exclusive fossil localities are now also beautifully illustrated on their website.

This brings me to the impact that the web has had on disseminating anything you wish to learn about trilobites. It now seems redundant for me to once again go over the detailed description of the trilobite's vital data and the intricacies of their paleontological classification. Instead, in this new book, I focus on aspects that are usually deemed irrelevant and thus glossed over. Who ever talks about the trilobites' aesthetic appeal, at times lyrical in extravagant shapes and patterns, or their often surprising color? Although the latter is generally an addition originating in the fossilization process, I will dwell in the following introduction on what is gained by viewing native color images of trilobites, as they appear when we first pry open their entombment.

1

INTRODUCTION: WHY COLOR?

As we split open a slab of shale, especially if it is of Cambrian age, trilobites often appear in brilliant shades of ochre, against a gray-greenish background, the color of the matrix encasing them. The color of the trilobite shell, or carapace, has little to do with the original color of the living trilobite. It is the result of chemical processes, such as mineralization, which took place after the burial and decay of the dead animal or its molting exuviae. These processes are called diagenesis and are part of a broader set of burial events that depend on the environment and fall under the comprehensive name of taphonomy. Taphonomy involves the nature of the minerals making up the shale and of those dissolved in the surrounding water or infiltrated after burial and fossilization. And yet, the color that our vision perceives has an emotional impact that makes our "treasure hunt" for trilobites both attractive and addictive.

Unfortunately, the rigor of the scientific paleontological literature has predominantly denied such emotional responses by erasing any trace of color with whitening coatings, which reduce the graphical reproduction of the subject matter to black-and-white photography. However, there is a reason for the whitening process: the aerosols used for the whitening carry an electrical charge that makes the coating stick preferentially to the tips of protruding details, thus enhancing somewhat the resulting informative three-dimensional contrast. Such whitening has been the fate of many images presented in my earlier trilobite books.

The rise of digital photography and the fact that color printing is now more affordable make it possible to exploit to its fullest the splendid complexity of human vision, which, with its trichromatic sensitivity, unique among primates, connects directly to the emotional response that color images may elicit as well as to artistic expression. Is this of added value over black-and-white images when looking at trilobites? It is my contention that, indeed, color images of trilobites convey more information about the fine structure of the object, such as that enhanced by aerosol whitening, than is usually appreciated. Under illumination of a color reproduction similar to that used in taking the picture (e.g., daylight or halogen lighting,), the response of the eye retina contains a subtle contribution from a phenomenon called chromostereopsis. The latter is an optical phenomenon whereby different colors may elicit different depth perceptions. Much has been discussed in the literature in regard to this phenomenon and how the brain processes the signals from the retina in regard to color processing hand in hand with the perception of form and boundaries. The net result is that color images seem to stand out with some illusion of relief. (An excellent and updated summary of this visual phenomenon can be found on the web under "Chromostereopsis" in Wikipedia, The Free Encyclopedia.)

My input on this issue is a comparison I made between black-and-white images of trilobites whitened with white aerosols of magnesium oxide (the old trick to enhance 3-D contrast mentioned above) and color images of the same prior to whitening (an example of this comparison is shown in plate 1). This comparison showed clearly that the structural information delivered by both images is essentially identical, thus confirming that the color image does provide the relief evidence sought by paleontologists for publication. On this ground, to publish my trilobite pictures in color has more to offer than just the entertainment of browsing through a coffee-table book.

Now I must turn to the choices I made about which trilobite images to include and which of their geographic origins to highlight. On both grounds, my choices were greatly biased. As to the choice of depositional location, this was dictated by preconceived notions of accessibility, renown, and sometimes also by exotic appeal. A certain predilection for the most ancient geological settings, the Lower and Middle Cambrian, will transpire here—a predilection perhaps fostered by my encounters

with my mentor, the late Franco Rasetti, who regarded Cambrian trilobites as the most rewarding of his attention. However, the principal bias results from my choice of illustrating complete, fully articulated trilobite specimens. Complete specimens are a rare occurrence in the fossil record of trilobites, which restricts greatly the taphonomy of their burial, which, in turn, is responsible for their ultimate appearance, inclusive of their color. Complete outstretched trilobites, with all sutures intact, indicate sudden burial of living individuals in a tranquil environment, at some depth in the sea, remote from the disturbance of wave motion. Such episodes of sudden burial are known to occur as a result of gentle but massive deposition of silt in suspension in so-called turbidity waves, the result of surface storms and sometimes volcanic ash fallout. The sediments often reveal a cyclic or rhythmic alternating repetition of mineral content, as documented, for example, in the layers of the Manuels River Formation of Conception Bay, Newfoundland—which I studied long ago with my colleague Jan Bergström of Sweden. After the death of a trilobite, several pathways lead to fossilization of the remains, which depend critically on the chemistry of the local environment. With few by now famous exceptions, like the Burgess

PLATE 1. This shows clearly that the structural information delivered by both images is essentially identical, thus confirming that the color image does provide the relief evidence sought by paleontologists for publication.

Shale of British Columbia and the Chengjiang fauna of China, where soft tissues are beautifully preserved, the decay of the biopolymers making up the soft tissues, generally fostered by bacterial action even in anaerobic marine sediments, leaves no trace of internal, nonmineralized structures. Only the trilobite exoskeleton, the carapace, partially mineralized by calcite, often escapes complete dissolution, and this may survive unaltered over geological eras. However, over time, mineralogical alterations, diagenesis, still conspire to modify the chemical nature of the fossilized carapace, leading to a multitude of final appearances, and in particular colors, of the fossil trilobite.

Thus, the brilliant ochre colors frequently encountered in the carapace of trilobites preserved in clay-mineral shales are the result of a three-step taphonomic-diagenetic process, favored by a marine environment rich in dissolved iron hydroxide and oxygen starved. In the first step, residual organic matter is digested by sulfate-reducing anaerobic bacteria yielding hydrogen sulfide (HS^-). In the second step, HS^- combines with iron ions (Fe^{2+}) to yield iron sulfide FeS_2 (the mineral pyrite, fool's gold). These two steps may occur within a few weeks after burial. The third step, which may occur over millions of years, involves the oxidation of the pyrite to the minerals limonite (FeO_2, yellow ochre) and hematite (FeO_3, red ochre). Sometimes, the pyrite replacement survives over eons, and we find the trilobites amidst a multitude of pyrite nodules, in particular in carbon-rich black shales.

Particularly striking yellow and red ochre carapace coatings are encountered in the Middle Cambrian *Paradoxides* of the Manuels River Formation of Newfoundland; Lower and Middle Cambrian, Ordovician, and Devonian of Morocco; the Ordovician of China (see plates 1 and 220); and the Lower Cambrian of California and Nevada. Very attractive coloring is found in the Olenelloid trilobites of the top Lower Cambrian layers at Ruin Wash, in the Chief Range of Nevada. Here the dark trilobite exoskeletons, replaced by chlorite (a clay mineral), are embedded in a wafer-thin medallion or halo of yellowish calcite, and the ensemble is further colored by scattered remnants of iron, sometimes manganese oxides, the result of a multistep taphonomy and diagenesis. This process was studied in detail by my University of Chicago colleague Mark Webster.

A different pathway to preservation of the exoskeletal cuticle, which occurred for the Middle Cambrian trilobites of the Wheeler Shale in Utah's House Range, leads to a black, somewhat shiny color. Here again the prevalent taphonomic environment seems to have prevented the complete bacterial digestion of the nonmineralized components of the cuticle, which were preserved as black kerogenized carbon films. In addition, due to the proximity of abundant calcium carbonate in the sediments, a peculiar growth of laminar calcite developed, extending for a few millimeters on the ventral side of the carapace, in the so-called cone-in-cone habit. These black "padded" trilobites, among them *Elrathia kingii* and *Asaphiscus wheeleri,* became the most collectible and sturdy mementos of an excursion to the Wheeler Amphitheater.

There are many other "taphofacies" that may determine the color appearance of the preserved trilobites, ranging from the extreme cases mentioned above to the total absence of color differentiation between the fossil and the sediment matrix. The latter presentation occurs when the entire body of the buried trilobite is dissolved by decay and the resulting cavity is filled by the same minerals making up the sediment. This leads to the presence of an external impression or mold of the original body surface and to an internal impression (steinkern) of the inside surface of the trilobite exoskeleton, if still present. The filling of either impression by sedimentary material forms a cast, which is a replica of the original body. At times, a thin gap is left between internal and external molds in the space that was occupied by the exoskeleton, the last part of the body to be dissolved. This void gap is sometimes encrusted with tiny crystals of calcite or dolomite in calcareous rocks. Color, if any, may result from minerals that may have infiltrated the void gap.

There is a notable exception to all of the above when focusing on the visual system of trilobites, the earliest known in the animal kingdom. Escaping decay and diagenesis, the lenses in the compound eyes of trilobites were made of oriented calcite crystals, already fossils in vivo. Such lenses are generally colorless and often still transparent. As described in detail in my previous trilobite books, some, like those of the Phacopids, had sophisticated and optimized doublet optics that corrected the chromatic aberration of thick spheroidal lenses. When we look at these eyes, we feel as if they were looking at us in their whimsical eternity.

2.1 BOHEMIA

The Bohemian Karst, a picturesque region southwest of Prague in the Czech Republic, has been immortalized by the blazing paleontological work of a French engineer, Joachim Barrande, who established himself in Prague in the middle of the 19th century. I first became familiar with his unparalleled scientific observations and descriptions of the trilobites of Bohemia when I perused the introductory volume of his series entitled *Système Silurien du Centre de la Bohème* among the dusty collections of the old geological library at the University of Chicago. I was struck by the magnificent lithographic plates of his descriptive documentation. And finally, in the mid-1990s, I found myself in the very locales of Barrande's discoveries in Prague and the Bohemian Karst, the latter now referred to as the Barrandian. Barrande's tangible legacy, his trilobite collection, is displayed at Prague's National Museum, where I spent many enraptured hours.

My visit to some of the historical Barrandian trilobite localities was greatly helped by the generous assistance of several scientists of the Geological Institute and the Czech Geological Survey. Among them, Petr Storch, Petr Budil, and the late Ivo Chlupác. In addition, Vratislav Kordule, an enthusiastic and knowledgeable amateur with an impressive trilobite collection and publication record, joined my quest on several occasions.

After absorbing the initial wonderment at the fairy-tale views of the Staré Město, Prague's Old Town, and becoming less inhibited at driving through the local traffic, I ventured toward my first field trip through the Bohemian Karst, kindly guided by Petr Storch. We exited a comfortable expressway at Beroun, and our first stop was Loděnice, where, climbing through a wooded slope, we reached the remnant of a famous Middle Silurian exposure, which in the past yielded the delightful *Odontopleura* and *Miraspis* trilobites shown in my previous books. These specimens were on loan from the Harvard Museum of Comparative Zoology and were part of the Agassiz Collection, acquired in the 19th century from a Bohemian source. Because they are irreplaceable, they will appear once more here, this time in color. What was left of the original excavations was only a thick layer of detritus, hiding the bedrock layers, and only a few small, yet beautiful, specimens of *Aulacopleura* could be obtained by splitting leftover chips.

Our trip continued, through attractive old villages and idyllic vistas, toward Jince, another historical locality noted for its Middle Cambrian *Paradoxides* beds. Above the Jince cemetery, a path through wildflower meadows brought us to a prolific *Paradoxides gracilis* locality, where I could extract a number of well-preserved exuviae. For me, it was a dream come true, as if Barrande's beautiful lithographs had come back to tangible reality in my hands. Somewhat further, in a pine forest, a quarry still yielded slabs containing assemblages of *Ellipsocephalus hoffi*, another abundant Middle Cambrian trilobite.

Appeased with the above introduction to the Barrandian, I was kindly introduced by Petr Storch to the scientists at the Czech Geological Survey (CGS) in Prague, located in an imposing building at the foot of the steep climb toward the Prague Castle and St. Vitus Cathedral. This building formerly housed a hospital retreat. After Petr Budil showed me several trilobite collections, he brought me to see the CGS library. To my surprise, I found myself on a balcony perched on the wall of a chapel adorned with beautiful frescos, remindful of the Sistine Chapel of the Vatican, overlooking a floor densely packed with book stacks, which had replaced the original pews. After absorbing this unexpected view, I was asked by Petr Budil if I wished to go for a quick trilobite hunt, another unexpected surprise, and he provided me with a geological hammer. I followed him for a short walk around the building, and found myself on a steep forest incline jutting on a rocky exposure, and we both started rummaging through the underlying detritus. Believe it or not, we were finding exuviae of Ordovician Trinucleid trilobites with pitted wide fringes in the middle of Prague!

The memories of my encounters at the CGS do not stop at this adventure. In a subsequent visit, accompanied by my colleague and longtime friend Mark Utlaut, which took place on the occasion of my 69th birthday, we were summoned to the office of the CGS director, Zdenek Kukal, together with Petr Budil and Ivo Chlupak, as well as several other CGS researchers. An imposing grand piano was part of a sumptuous decor. After the initial salutations, the director sat at the piano, and with his accompaniment, everybody erupted in a festive "Happy Birthday to You." It was an emotional moment.

My exploration of the Barrandian was further broadened with the generous help of Ivo Chlupak, who brought me and Mark Utlaut above Rejkovice on the way to Jince, up the mountain to a secluded exposure rich in *Paradoxides gracilis,* inclusive of several juvenile specimens, as well as *Conocoryphe sulzeri.* In a later comprehensive tour, we stopped at Skryje, where an imposing bronze likeness of Joachim Barrande commemorates his discoveries in the region, and went to dig on a nearby steep ravine where bright-yellow exuviae of *Hydrocephalus carens* could still be extracted from the exposed shale bedrock. We continued to Tyrovice, where another ravine yielded a collection of the remnants of more Middle Cambrian trilobites.

The next experience brought us to Pribram, where we met with Vratislav Kordule, who guided me to several further localities. Vratislav Kordule was instrumental in obtaining for me, Mark Utlaut, and a Milanese friend, Bruno Corti, permission from the local environmental protection authority to visit and collect at protected classical localities. In a memorable outing, an agent of the environmental protection authority guided us to a famous, now-hidden excavation at Vinice, at the periphery of Jince, on a densely forested right bank of the Litavska River. We were struck by the demure behavior of our guide, a burly, mustachioed, authoritative young man with a gentle touch, who kept checking wildflowers for evidence of parasites, while we traipsed through meadows leading to the river. After a barefoot crossing of the

Litavska, there we were, next to a wall of black rock that most likely yielded some of Barrande's discoveries. The rock wall was not as impervious as it looked, and sizeable blocks could be eased out for further splitting. Soon we augmented the underlying heap of prior chippings, amidst exclamations of success in finding shining bodies of *Paradoxides gracilis* and *Conocoryphe sulzeri.* The black sheen (probably due to surface diagenesis of the carapace into hematite) against the black matrix did represent a later challenge to my photography.

Vratislav Kordule guided me to another interesting locality, an immense quarry at Kosov, near Beroun, where Silurian strata were temporarily exposed. The strata contained very small but delightfully spiny *Acantholomina minuta,* which could be extracted, together with *Otarion diffractum* and other microfauna. Interestingly, these trilobites appeared dark when first exposed to the air from their wet environment, but turned an absolute white after drying. This transformation maximized the contrast between the object of interest and the dark background.

My excursions in the Barrandian were limited to collecting a very small fraction of the enormous variety of trilobites that were discovered and described by the grand master Joachim Barrande. It turns out that my collections from Morocco filled many gaps in my coverage: a large majority of the Moroccan species illustrated in this book, particularly those belonging to the "Pragian" period of the Devonian, correspond closely to their Bohemian counterparts, which are exhibited in their complexity in Prague's Barrandeum.

PLATE 2. *Conocoriphe sulzeri* Schlotheim (54 mm). M. Cambrian, Jince, *Paradoxides gracilis* Zone, Bohemia. Courtesy of MCZ, from the Agassiz Collection, now at FMNH. This trilobite was blind.

PLATE 3. *Conocoriphe sulzeri* Schlotheim (74 mm). Collected by the author in the same strata as for plate 2, exposed at Vinice near Jince, kindly guided by Vratislav Kordule.

PLATE 4. *Paradoxides gracilis* (Boeck) (13.0 cm). M. Cambrian, Vinice near Jince, *Paradoxides gracilis* Zone, Bohemia, RLS coll. Complete adult individual. The slab carrying this specimen had been subjected to linear tectonic shear at a skew angle of 21 degrees. The corresponding image distortion could be corrected for by digital processing.

PLATE 6. *Paradoxides gracilis* (Boeck) (13.1 cm vertical slab size). M. Cambrian, Jince, *Paradoxides gracilis* Zone, Bohemia. An assemblage of exuviae of similar size. RLS coll.

PLATE 7. *Paradoxides gracilis* (Boeck) (15 mm). M. Cambrian, Rejkovice near Jince, *Paradoxides gracilis* Zone, Bohemia. Late meraspid degree. Note elongated genal and second pleural spines. Collected by the author, kindly guided by Vratislav Kordule.

PLATE 5. *Paradoxides gracilis* (Boeck) (13.0 cm). M. Cambrian, Vinice near Jince, *Paradoxides gracilis* Zone, Bohemia. The librigena (free cheeks) have been shed in this carapace, the first stage of molting. RLS coll.

PLATE 9. Another example of exuviae of *Hydrocephalus carens* Barrande, as per plate 8 (16.3 cm). Also courtesy of George Ast.

PLATE 8. *Hydrocephalus carens* Barrande (14.7 cm). M. Cambrian, Jince Formation, *Eccaparadoxides pusillus* Zone, Skryje, Bohemia. Partially decomposed exuviae. Photographed by the author, through courtesy of George Ast.

PLATE 11. *Eccaparadoxides pusillus* (Barrande) (25 mm). M. Cambrian, Jince Formation, *Eccaparadoxides pusillus* Zone, Rejkovice near Jince, Bohemia. Partially enrolled. Gift to the author by Vratislav Kordule.

PLATE 10. *Hydrocephalus minor* (Boeck) (63 mm). M. Cambrian, Jince Formation, *Paradoxides gracilis* Zone, Vinice Hill near Jince, Bohemia. Incomplete exuviae. RLS coll.

PLATE 12. *Ellipsocephalus hoffi* (Schlotheim) (14.0 cm image longest side). M. Cambrian, Jince Formation, lowest *Paradoxides gracilis* Zone, Jince, Bohemia. Characteristically densely populated assemblage. Collected by the author, kindly guided by Petr Storch of the Geological Institute, Academy of Sciences of the Czech Republic.

PLATE 13. *Ptychoparia striata* (Emmrich) (72 mm). M. Cambrian, Jince Formation, *Paradoxides gracilis* Zone, Jince, Bohemia. Photographed by the author, through courtesy of George Ast.

PLATE 14. *Deanaspis goldfussi* (Barrande) (58 mm image longest side). Ordovician, Letná Formation, Letná Hill outcrop below Hanavsky Pavilion, Prague. A slab of cranidia and fragmentary exuviae. Collected by the author, kindly guided by Petr Budil of the Czech Geological Survey.

PLATE 15. *Deanaspis goldfussi* (Barrande) (19 mm). Ordovician, Letná Formation, Letná Hill outcrop below Hanavsky Pavilion, Prague. Collected by the author, kindly guided by Petr Budil of the Czech Geological Survey.

PLATE 16. *Otarion diffractum* Zenker (5 mm). M. Silurian, Kopanina Formation, Kosov quarry near Beroun. Collected by the author, kindly guided by Vratislav Kordule. Internal mold, coated with white calcium carbonate replacing the exoskeleton.

PLATE 19. *Miraspis mira* (Barrande) (14 mm). M. Silurian (Wenlock), Liteň Formation, Loděnice, Bohemia. RLS coll. Formerly from the Agassiz collection of MCZ, now at FM. White calcium carbonate replaces portions of the exoskeleton.

PLATE 20. *Miraspis mira* (Barrande) (25 mm). M. Silurian (Wenlock), LiteÐ Formation, Loděnice, Bohemia. RLS coll. Same origin, disposition, and remarks as for plate 19.

PLATE 17. *Acantholomina minuta* (Barrande) (5 mm). M. Silurian, Kopanina Formation, Kosov quarry near Beroun. Collected by the author, kindly guided by Vratislav Kordule. External (convex) impression of the mold, coated with white calcium carbonate replacing the exoskeleton.

PLATE 21. *Odontopleura ovata* Emmrich (28 mm). M. Silurian (Wenlock), Liteň Formation, Loděnice, Bohemia. Loaned courtesy of MCZ, from the Agassiz collection. External (concave) impression, carrying the exoskeleton replaced by calcium carbonate.

PLATE 22. *Odontopleura ovata* Emmrich (10 mm). M. Silurian (Wenlock), Liteň Formation, Loděnice, Bohemia. RLS coll. Same origin, disposition, and remarks as for plate 19. Two cephala of *Aulacopleura konincki konincki* (Barrande) are visible at the top of the photograph.

MOROCCO 2.2

My acquaintance with the trilobites of Morocco dates from the early 1990s, when they began to appear at the Tucson Gem and Mineral Show and, with much admiration on my part, in the collections of my surgeon friends David C. Rilling and Pio Pezzi. For the first time, these specimens had been expertly prepared by an Italian paleontologist, Flavio Bacchia. At that time, the fossil preparation by the early Moroccan diggers left much to be desired; it was usually carried out with primitive instruments, mostly nails, and much distorted imagination in the outlining of missing morphological details. I was finally lured to a firsthand encounter with pristine trilobites in 1996 with a two-week field trip. This was eventually followed by another 11 annual expeditions under the expert guidance of an enterprising young man I had originally met in Tucson, Abdullah Adam Aaronson, who had founded a tour company named Sahara Overland. These field trips usually encompassed visits to many localities at the southern edge of the Anti-Atlas range, at the northern edge of the Sahara desert, and have extended from the western edge of the Draa Valley, near the Atlantic coast, to the eastern edge of Morocco in the Tafilalt, near the Algerian border. Starting from Casablanca and ending in Rabat, we generally traveled some 3,000 kilometers on every tour.

The rich fossiliferous layers, exposed throughout in mountain ranges devoid of vegetation, span in geologic time from the Lower Cambrian to the end of the Devonian. The same pattern of layers may extend over many kilometers, marked by excavated trenches through preferred levels that the eye can follow to a distant horizon.

These exposures offer an ideal opportunity for trilobite collectors: they are easy to approach in most cases, with due precautions. Only the walls of the Grand Canyon offer a similar cross section of sediments deposited over some 200 million years. The layering of the sediments is not always orderly and tranquil. Geological turmoil is often evident and picturesque, with zigzag patterns, sometimes interrupted by volcanic thrusts and color changes. The strata sequences typically alternate over a variety of depositional environments, from hard limestones where the trilobites retain their full undistorted three-dimensional volume to softer shales where the trilobites may be vertically compressed. By and large, there is abundance of fossil localities where the majority of the trilobite-rich deposits yields specimens with minimal tectonic distortion.

Although the geological layout was always a cause of wonderment, other aspects of the landscape were equally fascinating, like the spring blooming of the almond trees in the mountain valleys, the refreshing palm trees shading an occasional oasis, goats climbing argan trees to feed on their olives, and many other travelogue-worthy escapades. The man-made contributions to the landscape were also welcome interruptions of our meandering, from the windowless casbahs in Berber villages dotting mountain slopes to the imposing archeological remnants of ancient native civilizations, as well as Roman, Portuguese, and French colonies. Thanks to the French occupants, we could enjoy baguettes with our morning meals, mostly north of the Atlas range. Our nights were usually spent in modern hotels with Arabian Nights decor, some of them memorably luxurious. Only once did we camp in the desert amidst scorpions, and only once on the floor carpets of a Berber home, with water-bucket facilities and no electricity. Regardless of the social level of our hosts, we always enjoyed the warmest hospitality. It was while in that Berber abode that one night I ventured in the dark outdoors and experienced the magical sight of the Sahara starlit sky, uncompromised by local illumination. It was an overwhelming presence, which made me aware of the reason why astronomy became so advanced in ancient Arabia. The extreme transparency of the moistureless desert atmosphere revealed to the naked eye a multitude of stars and nebulae that only telescopes can usually discern. This was an indelible memory renewed many times since.

The daily routine encompassed long morning desert drives, with some of our team preferring to ride on the roof, in search of a lonely shade tree, where we would stop for our picnic lunches. However, reaching some of the localities on our itineraries has at times been especially adventurous. Rainstorms in the sub-Sahara are usually accompanied by flash floods, whereby large, usually dry, riverbeds become tumultuous and frighteningly deep avalanches of unapproachable waters, capable of obliterating anything in their path. This happened on one of our tours, when, after three days of rainstorms, we found ourselves stranded in a village for a couple of days because bridges ahead and behind us had been swept away by the sudden downpours. Sandstorms have also frequently altered our schedule in springtime, usually lasting for three days or so. We once found ourselves blinded on the Paris-Dakkar desert path for several hours, having to resort to GPS navigation to reach our destination. The penetrating fine sand played havoc on our zoom cameras when we made the mistake of trying to take pictures of some of us wearing goggles and enveloped in turbans. Although the main roads are excellent, the trilobite hunts generally take place on Martian-like, off-road, bumpy or sandy terrain. Even our four-wheel-drive SUVs managed often to get stuck in the sand, and this is why a minimum of two vehicles was a must anywhere we went.

Besides trilobites, which require the most knowledgeable preparation unless found already exposed naturally, other abundant fossil assemblages make up part of a flourishing local industry. Large slabs densely populated by uncoiled nautiloids, Orthoceras, are sliced and polished to make table tops, and large, coiled Goniatites are typically polished to expose the intricate, delicate sutures that are pleasing to the eye. All this takes place in cramped ateliers with workers of all ages, amidst clouds of choking dust, as seen mostly in Erfoud, at the eastern edge of the Moroccan sub-Sahara.

My first encounters with the always friendly Berber trilobite diggers took place in humble abodes in the vicinity of Erfoud and Alnif, the trilobite capital of Morocco, and also under the tents of nomad families, amidst throngs of children asking for treats, bonbons, and stilos (ballpoint pens, a status symbol at school). Unaware of my already being a celebrity, I was instantly recognized and asked to autograph their copies of my trilobite books and astounded to see many of my trilobite pictures pasted on walls, even by analphabet admirers, as an aid to specific identification of trilobites made more valuable for sale. Even in the field, seemingly deserted as far as the eye could see from up a steep incline, children sometimes materialized out of nowhere, and once, one young man appeared carrying a bundle under his arm: lo and behold, it contained a copy of one of my books for autographing. Others brought me trilobites they had found while roaming about, hoping for a favorable exchange of gifts. From that day on, I always carried a supply of inexpensive watches and stilos.

Many of my finds were collected lying exposed in virgin territory, especially in dry rivulet beds, transported by rainwater. Many others were found in the debris of trenches dug out by local diggers, often valuable discarded external impressions. I often marveled at the ability of these humble trilobite searchers to recognize by name trilobite specimens whose existence in a split block of rock was only revealed by the thin and faint outline of the carapace cross sections. In such occurrences, the whole trilobite can be exposed by chipping away the enclosing matrix after reassembling and gluing together the two (or more) fragments of the split rock. Over the years, many times with my encouragement, a few local diggers became amazingly proficient in the use of pneumatic vibrotools and airbrasives for the superb preparation of intricate and delicate trilobite structures. Among the best apprentices of this skillful art is our friend Hmed, who now works in the atelier of our guide Abdullah Adam Aaronson.

Unfortunately, the advance in expertise on the part of some trilobite suppliers brought about artful repairs and faithful reproductions of truthful originals, which made the detection of forgeries difficult, unless done in the laboratory by forensic methods. This issue will be discussed further in the section devoted to the Tucson Gem and Mineral Show. The less damaging artifacts consist in the collage of authentic specimens from the same exposure into large and populated assemblies, masking the junctions between contiguous pieces with closely spaced nail scratches on the matrix surface. It is also unfortunate that the success of the fossil trade has spilled over to infect the native generosity of the local population. While in my early visits, the gifts I offered for the privilege of taking pictures of adorable children in colorful attire were shyly accepted by their mothers, now money is explicitly requested in advance for picture taking and permission is otherwise refused.

My quest for perfect, pristine specimens, which would qualify for my studio photography and possible inclusion in one of my books, took place not only by extracting and splitting promising rocks myself, but also in examining the prolific collections assembled by local diggers. Selection usually took place on floor displays at their homes. The homes typically consisted of one-flat constructions around an attractive patio, with colorfully painted metal doors giving access to carpeted living quarters, minimally illuminated by a tiny window or just by the opened door. This made the examination of their displays fruitful only by using flashlights, or more laboriously by taking each sample outdoors, between sips of the ever-present offer of mint tea. Sometimes, trilobite displays were set up at night, in the back trunk of four-wheelers, by collectors shy of publicity. Much bartering preceded each purchase in all cases. Often, local manpower was hired for help in digging out blocks of rock for my team to split, as well as for directing our search to the newest rich locations they had discovered. All in all, my quests were full of suspense and adventure, punctuated by unexpected findings. Many of the findings will be shown in this section about Morocco, split into chronologically ordered geological subsections covering most of the Paleozoic era.

CAMBRIAN

PLATE 23. *Daguinaspis ambroggii* Hupé (14 mm). L. Cambrian, Tazemmourt, Morocco. Amouslek Formation, *Daguinaspis* Zone. These early trilobites already had large eyes, as we can infer from the long palpebral lobes. This juvenile holaspid was covered by a thin calcitic layer and was photographed by the author as extracted from the sediment, without any preparation. RLS coll.

PLATE 24. *Daguinaspis ambroggii* Hupé (14 mm). L. Cambrian, Tazemmourt, Morocco. Amouslek Formation, *Daguinaspis* Zone. Another juvenile holaspid, completely exposed. Several cranidia of the same trilobite species are visible at the top, left side, and bottom of the image.

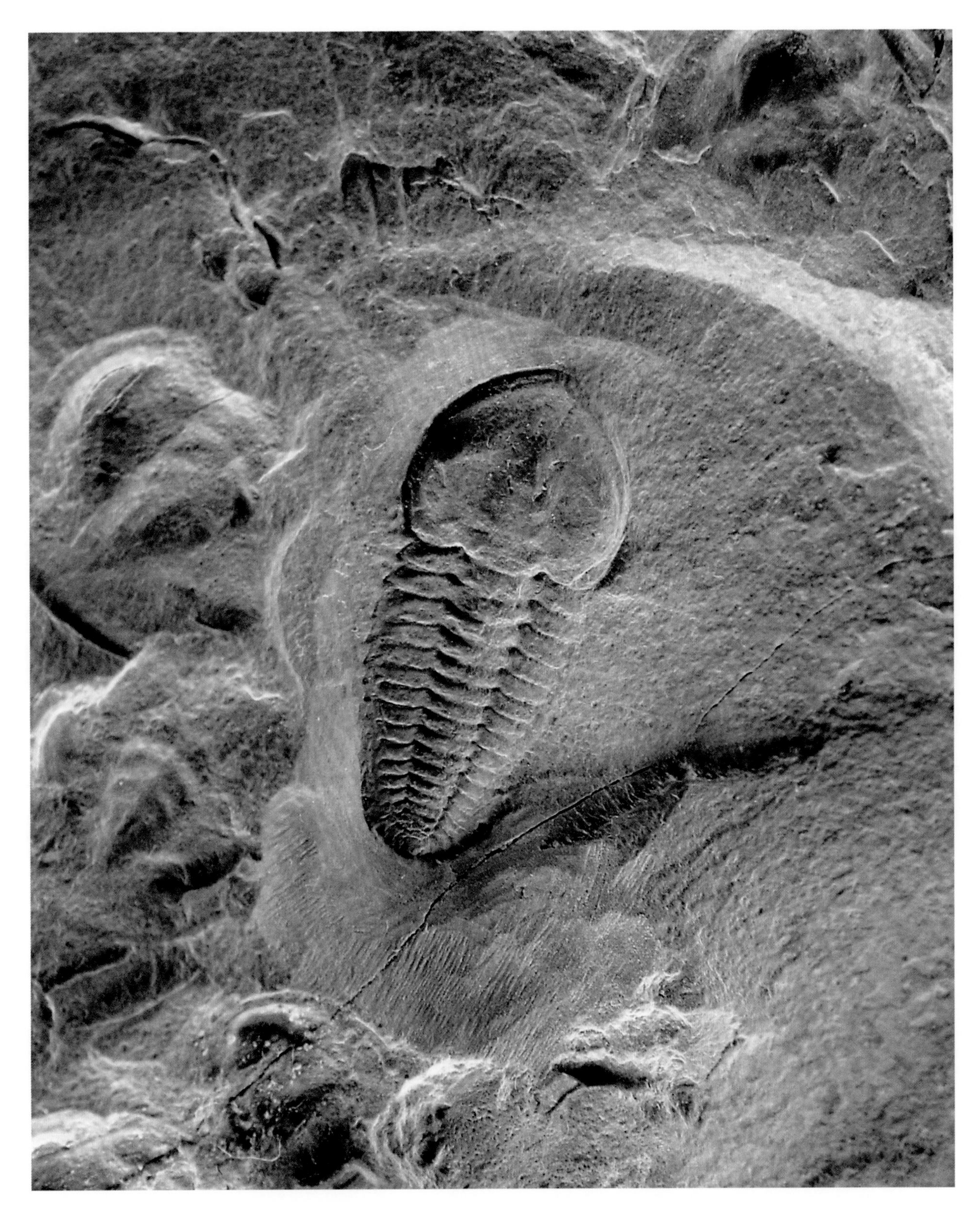

PLATE 26. *Daguinaspis ambroggii* Hupé (59 mm vertical image size). L. Cambrian, Tazemmourt, Morocco. Amouslek Formation, *Daguinaspis* Zone. Assemblage of four complete individuals. RLS coll.

PLATE 25. *Daguinaspis ambroggii* Hupé (34 mm). L. Cambrian, Tazemmourt, Morocco. Amouslek Formation, *Daguinaspis* Zone. This adult individual shows details of the arcuate short pleural spines. RLS coll.

PLATE 28. *Resserops (Richterops) falloti* Hupé (25 mm). L. Cambrian, Tazemmourt, Morocco. Amouslek Formation, *Daguinaspis* Zone. RLS coll. On a technical note, the excellent preservation of these two specimens describes visually their attribution to the new subgenus Richterops by Hupé (1953), based on the observation that the occipital ridge terminates above the occipital furrow.

PLATE 27. *Resserops (Richterops) falloti* Hupé (25 mm). L. Cambrian, Tazemmourt, Morocco. Amouslek Formation, *Daguinaspis* Zone. This specimen, together with the following image, represents a rare and much coveted finding of these neoredlichiidid trilobites from this locality. RLS coll.

PLATE 29. *Hamatolenus (Hamatolenus) vincenti* Geyer & Landing (9 mm each individual). Lower to Middle Cambrian boundary, Tarhoucht, Jbel Ougnate, Morocco. Jbel Wawrmast Formation, *Cephalopyge notabilis* Zone. RLS coll. Notable among these juvenile, complete holaspids is the long pleural spine emerging from the macropleural second thoracic segment in the two left-side individuals, the characteristic that best describes this species. The first thoracic segments in all the three trilobites appear macropleural as well, with shorter pleural spines.

PLATE 30. *Hamatolenus (Hamatolenus) vincenti* Geyer & Landing (43 and 20 mm). Lower to Middle Cambrian boundary, Tarhoucht, Jbel Ougnate, Morocco. Jbel Wawrmast Formation, *Cephalopyge notabilis* Zone. RLS coll. These larger, adult examples of this trilobite species are missing the librigena, and still show the first two anterior thoracic segments to be macropleural.

PLATE 32, 33. *Cambropallas telesto* Geyer (88 mm). Lower to Middle Cambrian boundary, Tarhoucht, Jbel Ougnate, Morocco. Jbel Wawrmast Formation, *Cephalopyge notabilis* Zone. RLS coll. These two mirror images represent the two sides of the matrix-encased trilobite, the cast (plate 32) and the mold (plate 33). In view of the explosive commercialization of these beautifully preserved fossils, leading to clever duplication, generally of the cast, the preservation of both sides of the impression is often telltale evidence of an original specimen, such as shown by these two corresponding images. The present identification of this trilobite supersedes my previous attribution of a similar specimen (see plate 62 of Levi-Setti, *Trilobites,* 2nd ed.) to the genus *Andalusiana,* indigenous of Spain. The bright-yellow color of these specimens is due to unadulterated coating by limonite (yellow ochre, FeO_2). The wide, backward-sloping cephalic shield of this trilobite suggests a hydroplanar function, facilitating swimming.

PLATE 31. *Hamatolenus (Hamatolenus) marocanus* (Neltner) (87 mm). Lower to Middle Cambrian boundary, Tarhoucht, Jbel Ougnate, Morocco. Jbel Wawrmast Formation, *Cephalopyge notabilis* Zone. RLS coll. Illumination from the south in this image allows a better definition of the anterior (preglabellar) border, which differs from that of *Hamatolenus (Hamatolenus) vincenti* Geyer & Landing. The large size of this trilobite is also a characteristic of this species.

PLATE 34. *Cambropallas telesto* Geyer (261 mm). Lower to Middle Cambrian boundary, Tarhoucht, Jbel Ougnate, Morocco. Jbel Wawrmast Formation, *Cephalopyge notabilis* Zone. RLS coll. A giant example of this trilobite, slightly distorted by tectonic shear. In this case, the red coating is due to red ochre (FeO_3).

PLATE 35. *Acadoparadoxides* sp. (14 mm). Lower Middle Cambrian, Bou Tiouit near Tarhia and Taroucht. Jbel Wawrmast Formation, Brèche à Micmacca Member. RLS coll. This and the following three plates describe examples of Paradoxidids that are provisionally assigned to the genus *Acadoparadoxides,* following discussions by Geyer (1993, 1998). A sequence of individuals of increasing sizes is shown, encompassing characteristics encountered in different subgenera of Bohemian-like *Paradoxides.* From a detailed stratigraphic study by Anthony Vincent (2010), the four examples selected here originate from layers encompassing the earliest appearance of *Paradoxides* in Morocco. This plate shows a meraspid with long eye ridges and macropleural second thoracic segment leading to a long pleural spine. The cranidial furrows are deeply impressed.

PLATE 37. *Acadoparadoxides* sp. (66 mm). Possibly *Acadoparadoxides mureroensis* Sdzuy. Same origin as that of plate 35. RLS coll. This increasing size sequence shows a progressive shortening of the ocular ridge and gradual medial reduction of the depth of the cranidial furrows.

PLATE 36. *Acadoparadoxides* sp. (34 mm). Possibly *Acadoparadoxides nobilis* Geyer. Same origin as that of the preceding plate. RLS coll. This is a fully developed holaspid with a well-defined pygidium, subtrapezoidal with truncated posterolateral angles.

PLATE 38. *Acadoparadoxides levi-settii* n. sp. Geyer & Vincent (190 mm). Same origin as that of plate 35. RLS coll. In a recent study (Geyer, private communication), these authors have assigned this specimen to be a paratype of a new species carrying my name. This is the largest individual in my collection, among a number that I dug out over many yearly field trips and whose images had been previously shown to Geyer for help in their identification. The new species attribution of this specimen may also supersede the previous tentative one given above for the smaller individual in plate 36.

PLATE 39. *Acadoparadoxides briareus* Geyer (327 mm). Lower to Middle Cambrian boundary, Tarhoucht, Jbel Ougnate, Morocco. Jbel Wawrmast Formation, *Cephalopyge notabilis* Zone. RLS coll. This is the largest of all Moroccan trilobites, often found in association with *Cambropallas telesto* Geyer, in layers above those of the Paradoxidids of the series of plates 35 to 38. For this trilobite, the ocular ridge is shorter than that observed in the preceding series, and the pygidium not as wide.

PLATE 41. *Parasolenopleura* sp. aff. *Parasolenopleura aculeata* (Angelin) (average size 48 mm). M. Cambrian prolific outcrop at Foum Lahssen near Tissint. RLS coll.

PLATE 40. *Acadoparadoxides briareus* Geyer (300 mm). This complete specimen overlapped another complete individual, partially exposed here. Found by a local digger named Moujan at Timarzit, it was photographed by Pat Sword, the professional photographer of my party, near the end of our 2009 field trip to Morocco. Regrettably, our exhausted finances at that time did not allow its purchase.

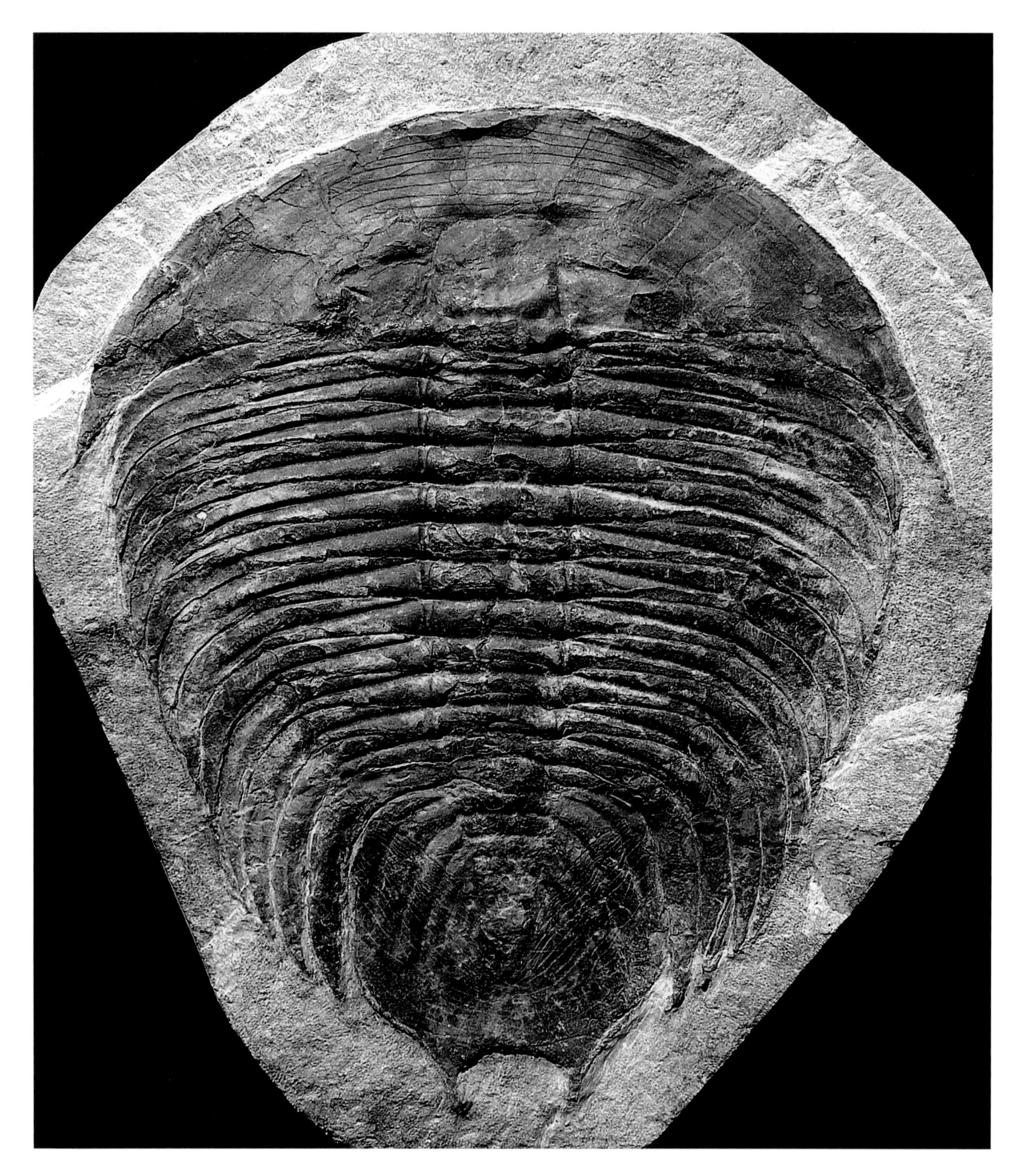

PLATE 43. *Diacalymene ouzregui* Destombes (each 63 mm). L. Ordovician, Ktaoua Formation, Tazoulait, Morocco. RLS coll. These trilobites are found encased in rounded mudstone nodules easily removed from a soft matrix. The nodules split open from a gentle percussion, revealing perfectly preserved casts, brightly coated by red ochre, that do not require any preparation.

ORDOVICIAN

PLATE 42. *Dikelokephalina brechleyi* Fortey (270 mm). L. Ordovician, L. Fezouata Formation, Ouled-Slimane, Ain Chika, near Agdz, Morocco. RLS coll. This large trilobite, previously illustrated in plate 141 of the second edition of my trilobite book (Levi-Setti 1993) as *Dikelokephalina* sp., has been recently described by Richard Fortey (2011), who named this genus species. Large and densely populated assemblages of this trilobite (see section on **Tucson Gem and Mineral Show**) suggest mass gatherings of mature individuals to fertilize deposited eggs, as observed for the living horseshoe crab, *Limulus*.

PLATE 44. *Symphysurus* sp. aff. *Symphysurus palpebrosus* (Dalman) (each 42 to 50 mm). L. Ordovician, near Alnif, Morocco. Known from Sweden.

PLATE 45. *Selenopeltis buchii* (Barrande) (16.3 cm longest sagittal length). U. Ordovician, Tiouririne Formation (Ktaoua Group) near Erfoud, Morocco. RLS coll. The alignment of these three trilobites suggests burial of living individuals oriented against water current. Preserved in very hard, mineralized sandstone, which required laborious preparation (by the author) using diamond-tipped tools under the microscope.

PLATE 46. *Selenopeltis buchii* (Barrande) U. Ordovician, Tiouririne Formation (Ktaoua Group) near Erfoud, Morocco. A densely populated assemblage, part of a meter-sized slab, in the process of being prepared in a local atelier. Photographed by the author in sunlight.

PLATE 47. *Selenopeltis buchii* (Barrande) U. Ordovician, Ktaoua Formation near Mcissi, Morocco. A view of the seafloor, the natural habitat where these trilobites roamed, surrounded by starfish and a population of brittlestars. These ophiuroids, still living, share with the trilobites a remarkable survival apparatus. Their skin is covered by a tapetum of microlenses made of oriented calcite crystals, shaped like the focusing-optimized lenses of the schizochroal trilobites, part of a photoreceptor system (Alzenberg et al. 2001). This large slab was photographed by the author while being prepared by adept young workers in Erfoud.

PLATE 48. *Selenopeltis longispinus* Vela & Corbacho (53 mm). U. Ordovician, Bou Nemrou, El Kaid Errami, Izegguirene Formation, Morocco. RLS coll. This trilobite, which differs from the well-known *Selenopeltis buchii* (Barrande) by two extra-long pleural spines of the eighth thoracic segment, appeared in the trilobite market several years ago. It was recently described (Vela and Corbacho 2009). After seeing several identical reproduction at the Tucson Show, I became suspicious as to the authenticity of the extra-long pair of pleural spines and have investigated possible fraudulent origins. The 2009 paper by Vela and Corbacho was helpful in dispelling my doubts.

PLATE 49. *Onnia superba* Bancroft (average length 14 mm). U. Ordovician, Ktaoua Formation. Bordj, Morocco. RLS coll. A typical assemblage of this trilobite, richly illustrated in the web trade.

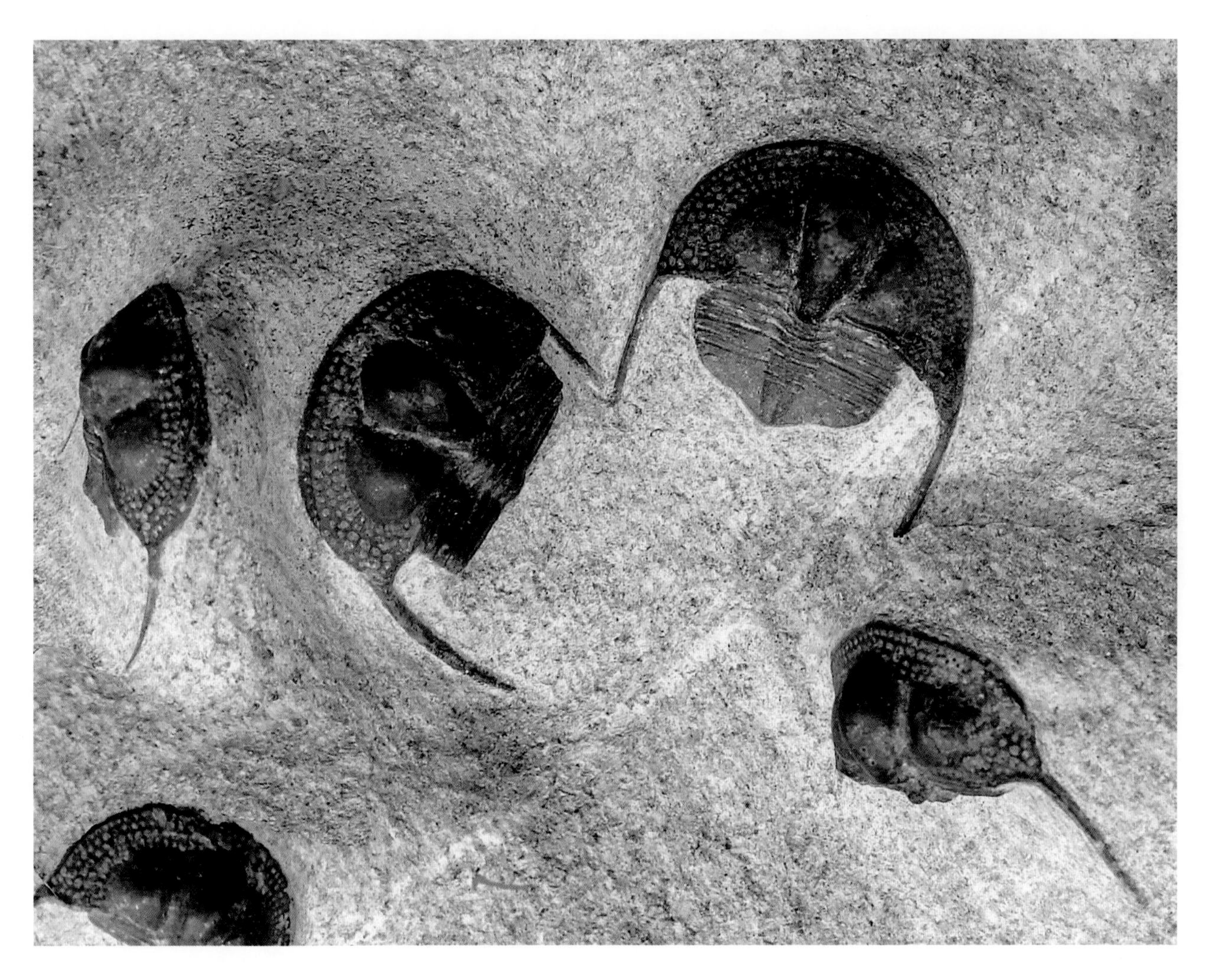

PLATE 50. *Onnia superba* Bancroft. A detailed view of several individuals, seen in a prep atelier in Erfoud.

PLATE 51. *Nankinolithus* sp. (31 mm). External impression, U. Ordovician, Bou Nemrou, El Kaid Errami, Izegguirene Formation, Morocco. RLS coll. This beautiful Trinucleid trilobite has attracted my interest for a long time. It is erroneously attributed to *Nankinolithus nankinensis* Lu in the trilobite trade, as I concluded from an exhaustive investigation aided by one of my Chinese physics students, Yu-Lin Wang, who translated for me the original 1954 publication by Yan-Hao Lu in Academia Sinica (Lu 1957). The original illustrations by Lu show a fringe that expands outward and terminates just below the cranidium, while in the Moroccan species it tapers gradually to extend well below the pygidium and terminates in a short genal spine as shown here. We are dealing with a new species, whose diagnosis is beyond the present scope. How I retrieved the present specimen from the field is unusual. It was found, together with several others, among the debris of an excavation for a water well, which, at some depth, crossed Upper Ordovician sediments near Tarhia, south of Tinejdad, Morocco. This locality is the typical source of abundant specimens.

PLATE 52. *Nankinolithus* sp. (43 mm width). Cranidium and fringe, from the same well sediments mentioned for the preceding plate. In technical terms, this plate provides details of the architecture of the girder (brim) of the fringe. Intermittedly, the upper and lower lamellae of this structure are exposed.

PLATE 54. *Dalmanitina* sp. aff. *Dalmanitina socialis* (Barrande) (24 mm). M.-L. Ordovician, Alnif, Morocco. RLS coll. Known from the Letná Formation, in the Barrandian of Bohemia.

PLATE 53. African reliquary figure (Bakota Tribe, Gabon), the Toledo Museum of Art. It has been my contention that primitive art forms may have been inspired by nature finds, in this case Trinucleid trilobites. Several characters of this adornment evoke details found in the preceding plates.

PLATE 55. *Ormathops* sp. aff. *Ormathops (Ormathops) atavus* (Barrande) (43 mm). M.-L. Ordovician, Alnif, Morocco. RLS coll. Known from the Šárka Formation, in the Barrandian of Bohemia.

PLATE 56. *Placoparia* sp. aff . *Placoparia (Placoparia) barrandei* (Prantl et šnajdr) (53 mm). M. Ordovician, Alnif, Morocco. RLS coll. Known from the Šárka Formation, in the Barrandian of Bohemia.

PLATE 58. *Phacops (Drotops) armatus* (Struve) (14.5 cm extended sagittal length). M. Devonian, South of Jbel Issoumour, near Alnif, Morocco. RLS coll. Plate 203 of my 1993 trilobite book anticipated the naming of this new species, prior to the development of the art of preserving spines. In that plate, the spines were truncated, but the stumps were still lined up in order. The specimen shown here is one of the first to appear with the spines, expertly prepared by a friend named Hmed in Erfoud.

DEVONIAN

PLATE 57. *Phacops (Drotops) megalomanicus* (Struve) (13.5 cm). M. Devonian, Jbel Issoumour, near Alnif, Morocco. RLS coll. After a steep climb to reach an excavation trench halfway to the summit, while carrying a sledgehammer, I started pounding on massive chunks of hard rock, and this treasure suddenly appeared. Touched by its aesthetic appeal, I had to preserve it as such, without further prep. This trench, extending around the mountain as far as the eye could follow, has been the source of a flood of specimens of this trilobite, which nowadays appears in rock shops worldwide and on the web.

PLATE 59. Another view of the head of the preceding specimen, to emphasize the appearance of the eye, with its spiny eyelashes.

PLATE 60. *Phacops (Drotops) armatus* Struve (life-size image). M. Devonian, South of Jbel Issoumour, near Alnif, Morocco. Courtesy of Bill Barker. An extravagant duo, in defensive posture.

PLATE 62. *Odontochile (Spinodontochile) spinifera spinifera* (Barrande) (71 mm). L. Devonian, Pragian, Hamar Laghdad Formation, lowest Jbel Issoumour levels near Tabourigt, Morocco. RLS coll.

PLATE 61. *Odontochile (O.) hausmanni* (Brongniart) (12.2 cm). L. Devonian, Pragian, Hamar Laghdad Formation, lowest Jbel Issoumour levels near Tabourigt, Morocco. RLS coll. It has been a ritual for me to visit this locality over many years and collect innumerable attractive pygidia of this trilobite.

PLATE 63. *Crotalocephalus (C.) gibbus* (Beyrich) (71 mm). L. Devonian, Pragian, south end of Jbel Issoumour, Morocco. RLS coll. These trilobites, often in association with *Odontochile,* are common in long trenches excavated at ground level.

PLATE 64. *Paralejurus dormitzery rehamnanus* Alberti (72 mm). L. Devonian, Pragian, Jbel Issoumour, El Oftal Formation, "Paralejurus beds," Morocco. RLS coll. Notable here is the ornamentation of terrace lines on the cephalon and the vaulted pygidium.

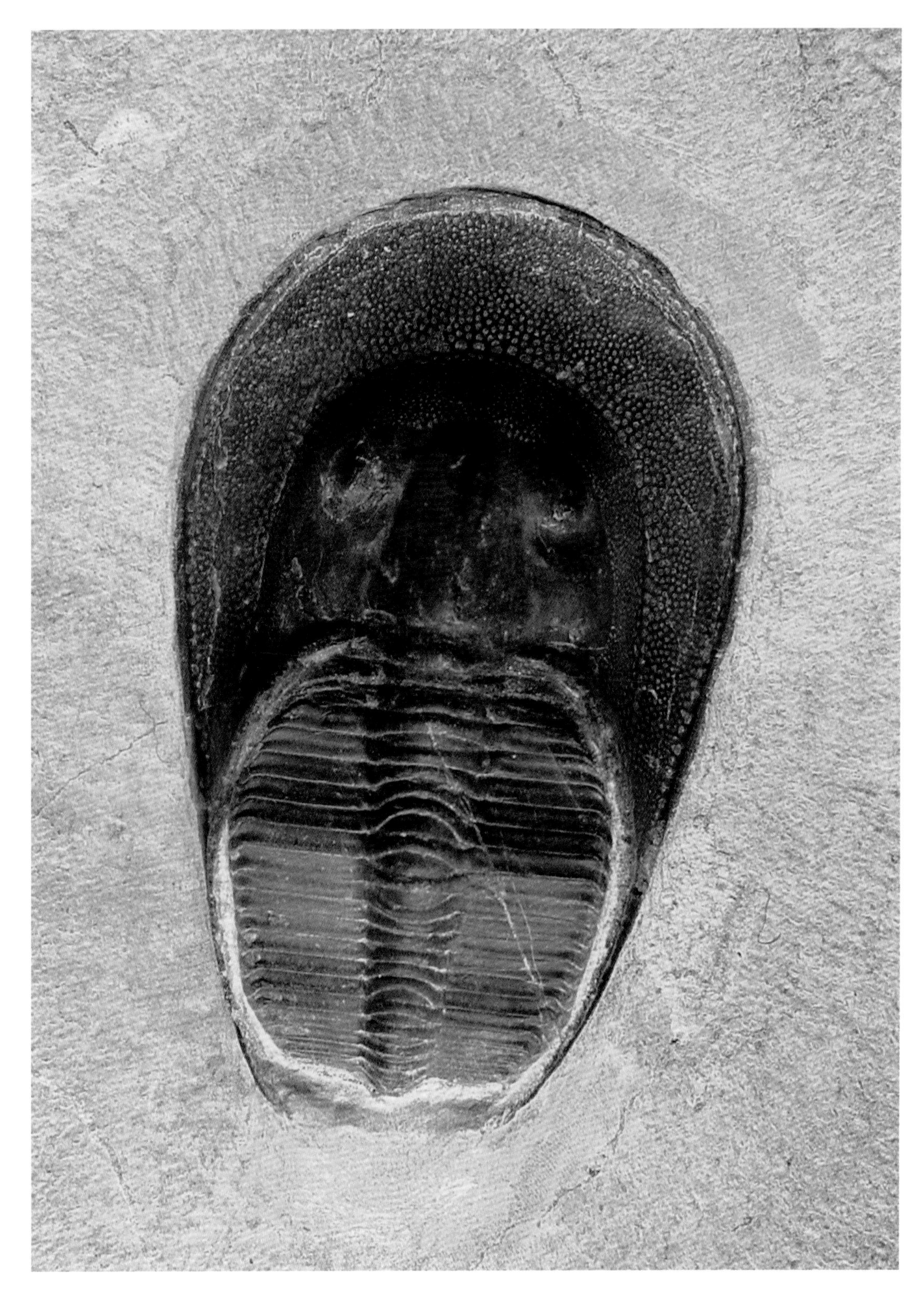

PLATE 66. *Acanthopyge* sp. (74 mm). L. Devonian, Atchana, Ma'der, south of Jbel Issoumour. Courtesy of Lang's Fossils. This is an elusive trilobite, rare in unadulterated complete state. Most exposures at this locality are by now thoroughly dug out and only fragments can be collected.

PLATE 65. *Harpes* sp. aff. *Lioharpes venulosus* (Hawle & Corda) (40 mm). L. Devonian, Atchana, Ma'der, south of Jbel Issoumour. Another reminder of the Barrandian of Bohemia.

PLATE 67. *Thysanopeltis speciosa* (Hawle & Corda) (48 mm). L. Devonian, Pragian, Hamar Laghdad Formation, Ma'der, south of Jbel Issoumour, Morocco. RLS coll.

PLATE 68. *Scutellum pustulatum* (Barrande) (5.2 cm). L. Devonian, Pragian, Hamar Laghdad Formation, Ma'der, south of Jbel Issoumour, Morocco. RLS coll.

PLATE 70. *Scabriscutellum (Cavetia) furciferum hamlaghdadianum* Alberti (slab size 29 cm ✢ 42 cm). L. Devonian, Pragian, Hamar Laghdad Formation, Hamar Laghdad, near Erfoud. RLS coll. This imposing slab of disarticulated carapaces was cut out for me by a marble cutter, from a massive block that I spotted in his marble yard in Erfoud. It originates from the type locality where a major exposure is laid out on top of a vertiginous cliff. I climbed this cliff several times to collect many densely populated rock pieces in my search of the beautiful holochroal eyes of this trilobite.

PLATE 69. *Scabriscutellum (Cavetia) furciferum hamlaghdadianum* Alberti (48 mm). L. Devonian, Pragian, Hamar Laghdad Formation, Jbel Oufatène, south of Jbel Issoumour, Morocco. RLS coll.

PLATE 71. *Cornuproetus (Diademaproetus) antatlasius* Alberti (34 mm). M. Devonian, "Barre de calcaire grèseuse, zone à *Pinacites*," Assa, Morocco. RLS coll.

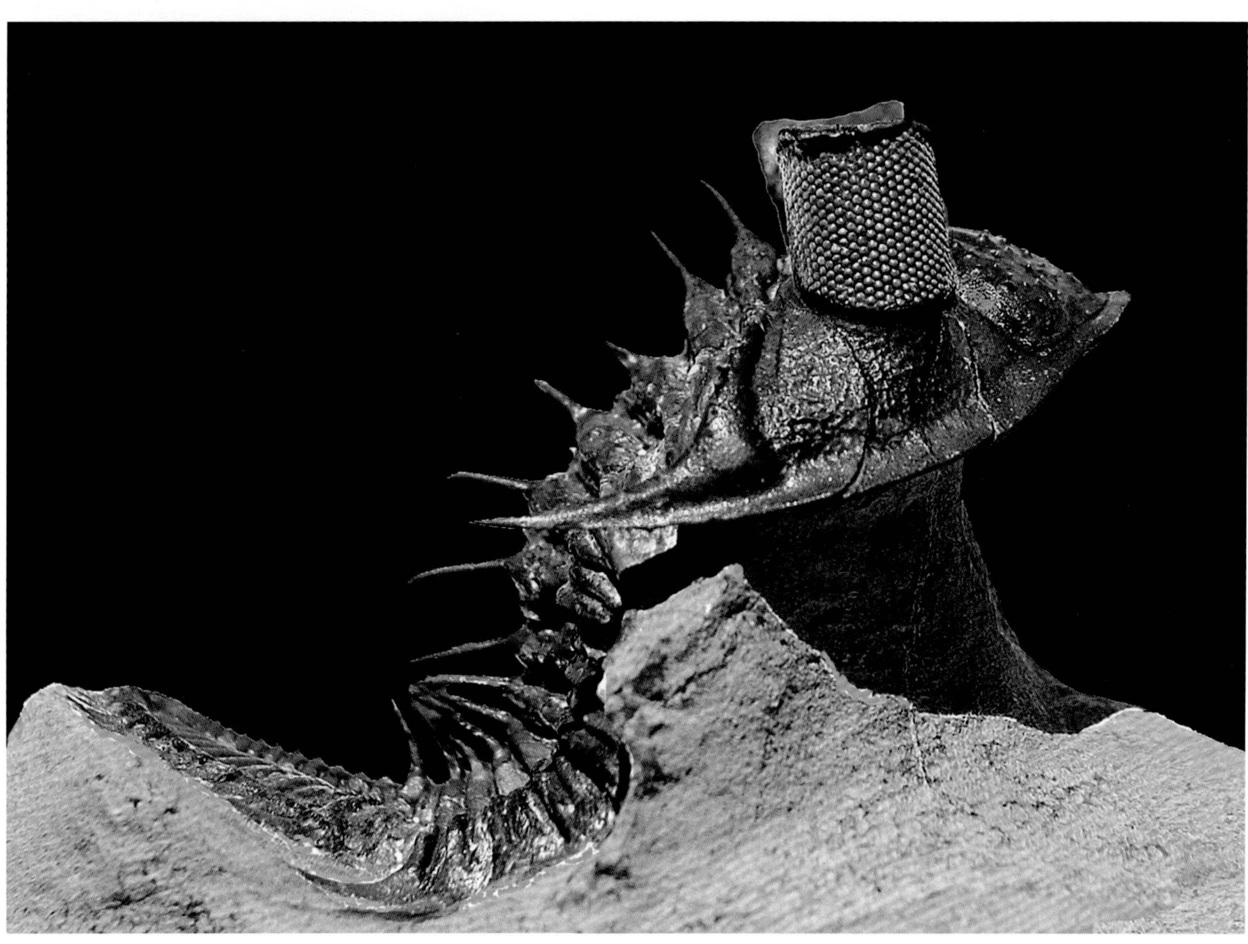

PLATE 72. *Erbenochile erbeni* (Alberti) (extended sagittal length 44 mm). L. Devonian (Emsian), Timrarhart Formation, near Foum Zguid, Morocco. Specimen courtesy of Adam A. Aaronson, photographed by Pat Sword. I will come back to this trilobite in the section on the eyes of trilobites to discuss the function of its amazing overhanging eyeshades.

PLATE 73. *Hollardops mesocristata* (Le Maître). L. Devonian (Emsian), Timrarhart Formation, Jbel Anhsour, south of Foum Zguid, Morocco. I photographed this partially exfoliated, complete carapace as shown here in the field while approaching a major excavation at this locality.

PLATE 74. *Hollardops mesocristata* (Le Maître) (57 mm). L. Devonian (Emsian), Timrarhart Formation, Jbel Anhsour, south of Foum Zguid, Morocco. RLS coll. Prepared by the author, photographed with ring lighting.

PLATE 75. *Metacanthina maderensis* Morzadec (62 mm). L. Devonian, Pragian, south of Jbel Issoumour, Ma'der, Morocco. RLS coll.

PLATE 77. *Coltraneia oufatenensis* Morzadec (62 mm). L. Devonian (Emsian), El Otfal Formation, Oufatène, south of Jbel Issoumour, Ma'der, Morocco. A close-up of the exceptional schizochroal eyes of this trilobite appears in the section on the eyes of trilobites.

PLATE 76. *Treveropyge maura* Morzadec (67 mm). L. Devonian (Emsian), Oufatène, south of Jbel Issoumour, Ma'der, Morocco. RLS coll.

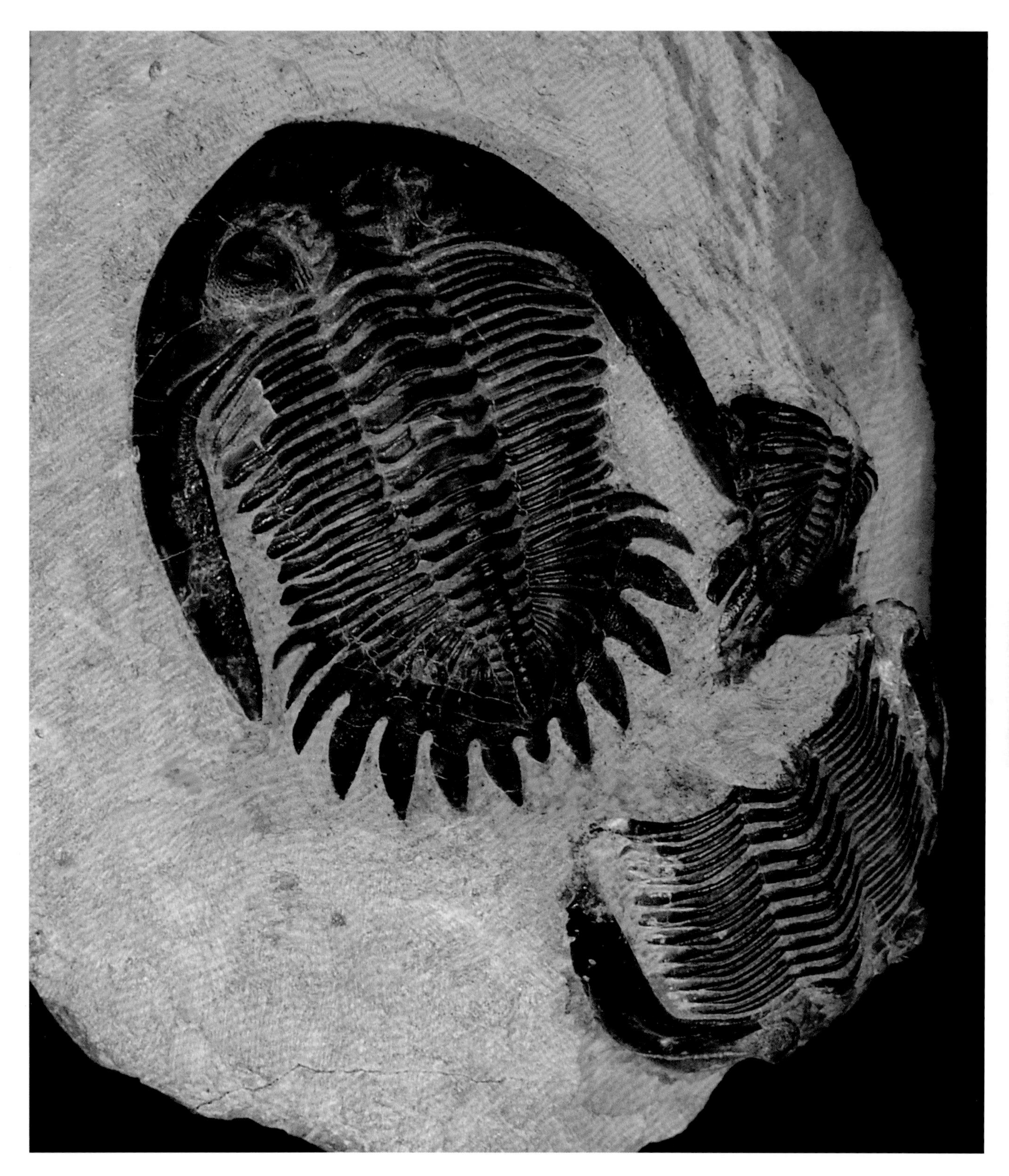

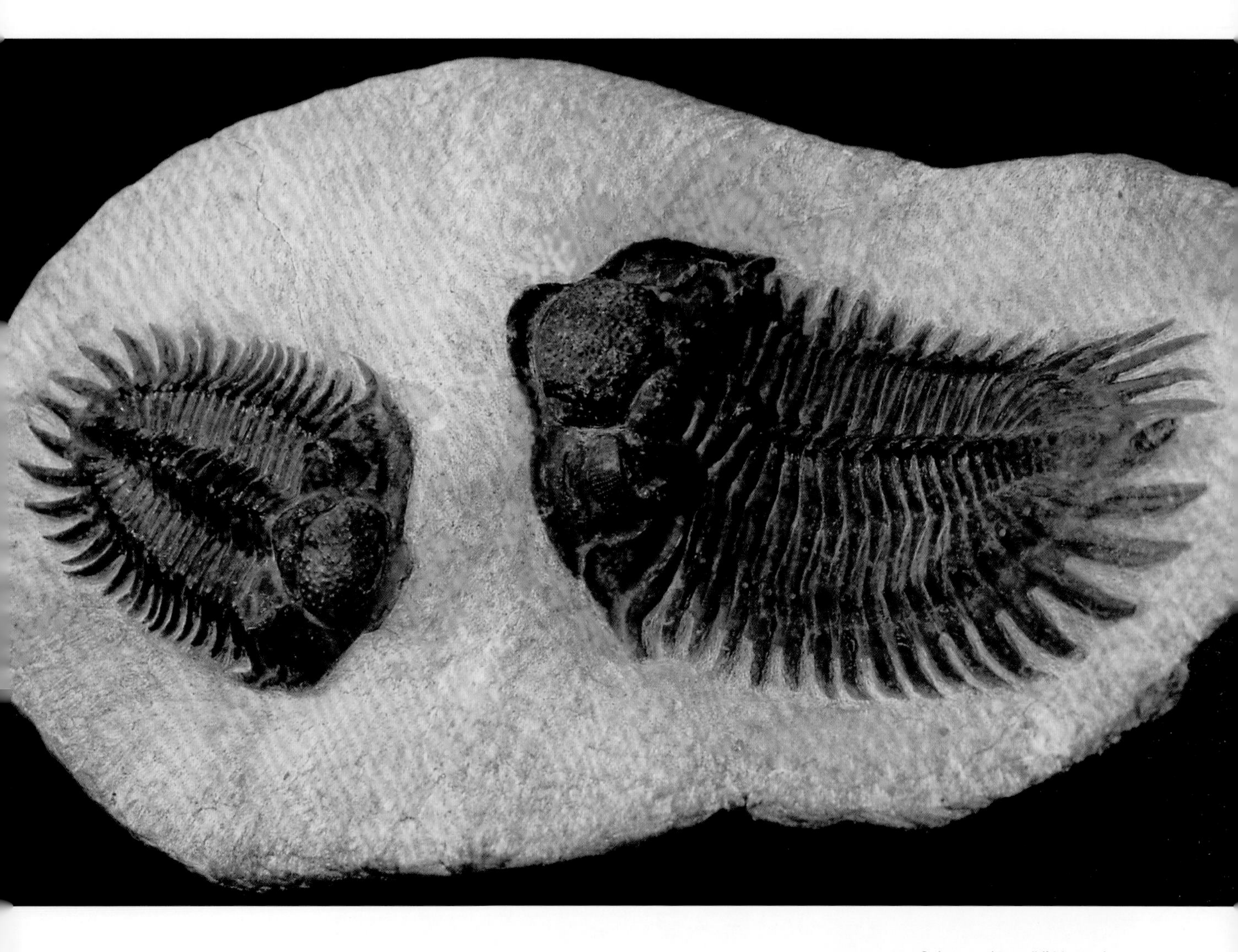

PLATE 79. *Saharops bensaïdi* Morzadec (largest specimen 78 mm). L. Devonian (Emsian), south of Jbel Issoumour, Ma'der, Morocco. RLS coll.

PLATE 78. *Mrakibina cattoi* Morzadec (36 mm). L.-M. Devonian, El Otfal Formation, Madène El Mrakib, South Ma'der, Morocco. RLS coll.

PLATE 80. *Metacanthina* sp. aff. *Metacanthina issoumourensis* Morzadec (74 mm). Devonian, far away from Jbel Issoumour, this trilobite originates from an unnamed outcrop near Rabat, Morocco. Telltale characteristic is the long terminal pygidial rachis, in relief and faintly ringed. Specimen courtesy of Adam A. Aaronson, photographed by the author.

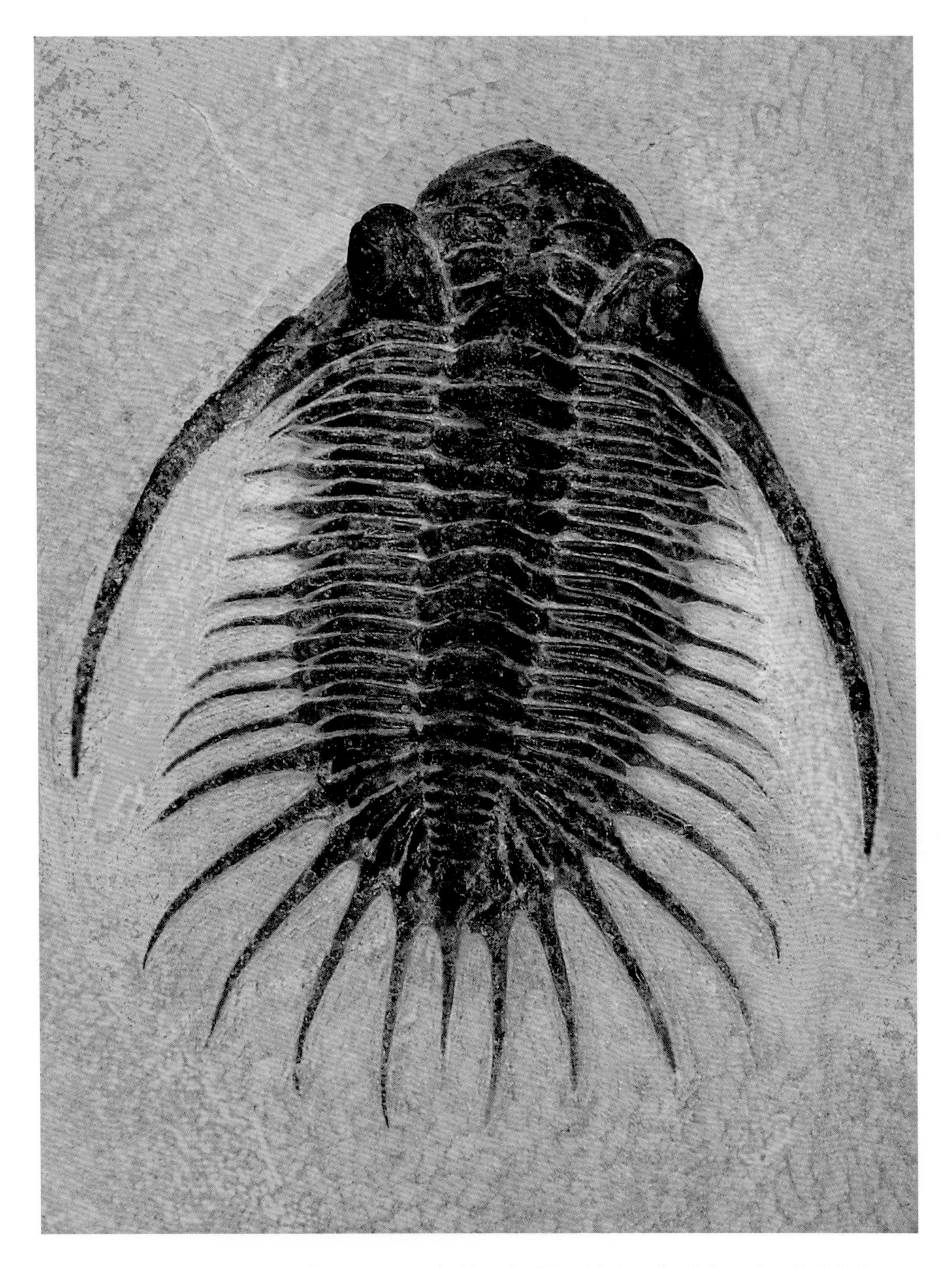

PLATE 81. *Kayserops megaspina* Morzadec (58 mm). L. Devonian (U. Emsian), south of Jbel Issoumour, Ma'der, Morocco. RLS coll. The surface of the carapace is granulose, and tubercles protrude along the median of the thoracic rings, as described by Morzadec (2001).

PLATE 82. *Comura bultincki* Morzadec (65 mm). L. Devonian (U. Emsian), Tazoulait Formation, Oufatène, Ma'der, south of Jbel Issoumour, Morocco. Specimen courtesy of Adam A. Aaronson, photographed by the author.

PLATE 83. *Psychopyge elegans* G. & H. Termier (13.8 cm). L. Devonian (U. Emsian), Tazoulait Formation, Jbel Oufatène, Ma'der, south of Jbel Issoumour, Morocco. RLS coll. I also collected several pygidia of this trilobite in my diggings at Jbel Anhsour, south of Foum Zguid. This specimen was expertly prepared by my friend Robert Carroll, who was able to extract it from a split block of rock, guided only by a filiform cross section of the carapace.

PLATE 84. *Walliserops trifurcatus* Morzadec (43 mm). L. Devonian (U. Emsian), Timrarhart Formation, Jbel El Gara, southwest of Foum Zguid, Morocco. RLS coll.

PLATE 87. *Dicranurus monstrosus* (Barrande) (44 mm). L. Devonian (Pragian), Atchana, Oufatène, Ma'der, south of Jbel Issoumour, Morocco. Specimen courtesy of Moussa Minerals and Fossils, photographed by the author.

PLATE 85. *Quadrops flexuosa* Morzadec, also known as *Philonyx philonyx* Richter & Richter (76 mm). L. Devonian (U. Emsian), El Otfal Formation, Oufatène, Ma'der, south of Jbel Issoumour, Morocco. Specimen courtesy of Bill Barker, photographed by the author. This trilobite, as well as the two preceding Asteropyginae, are provided with anterior cephalic processes that may have developed as tools to sift through detritus in search of food.

PLATE 86. *Koneprusia mediaspina* Alberti (43 cm). M. Devonian (L. Eifelian), Jbel Oufatène, Ma'der, south of Jbel Issoumour, Morocco. The advent of airbrasive techniques has enabled the preservation of the profusion of long spines found in this and other Odontopleurid trilobites.

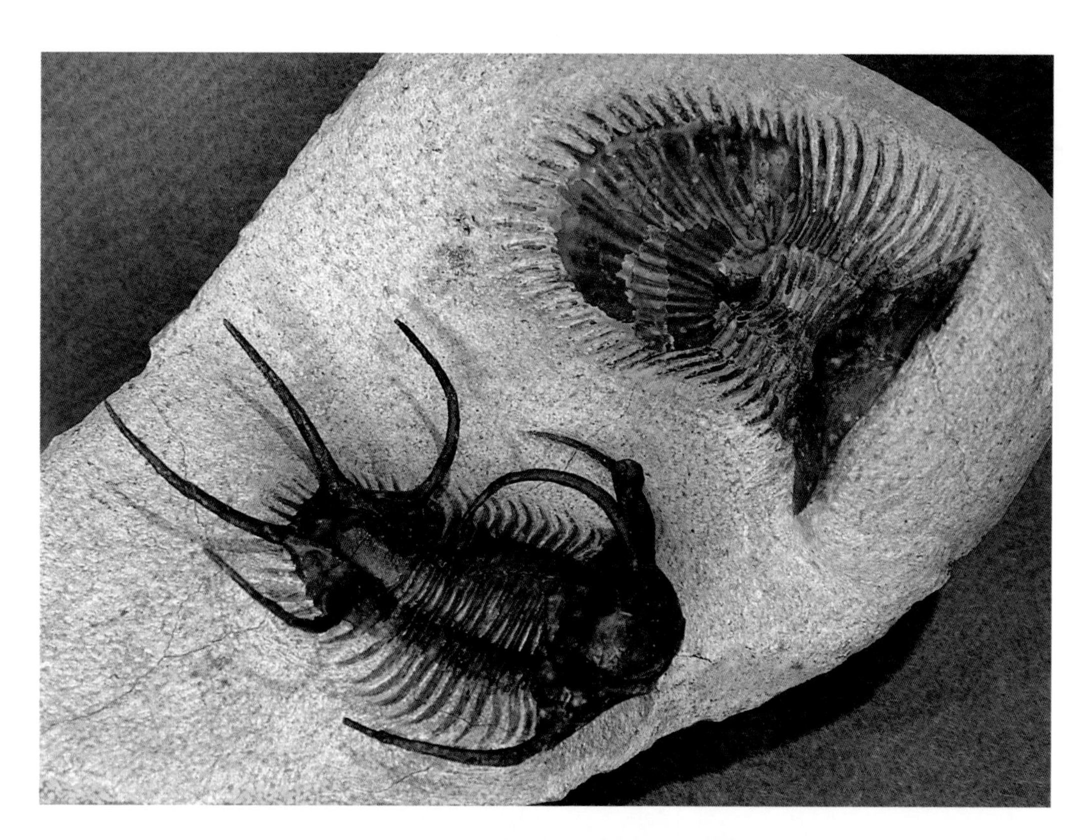

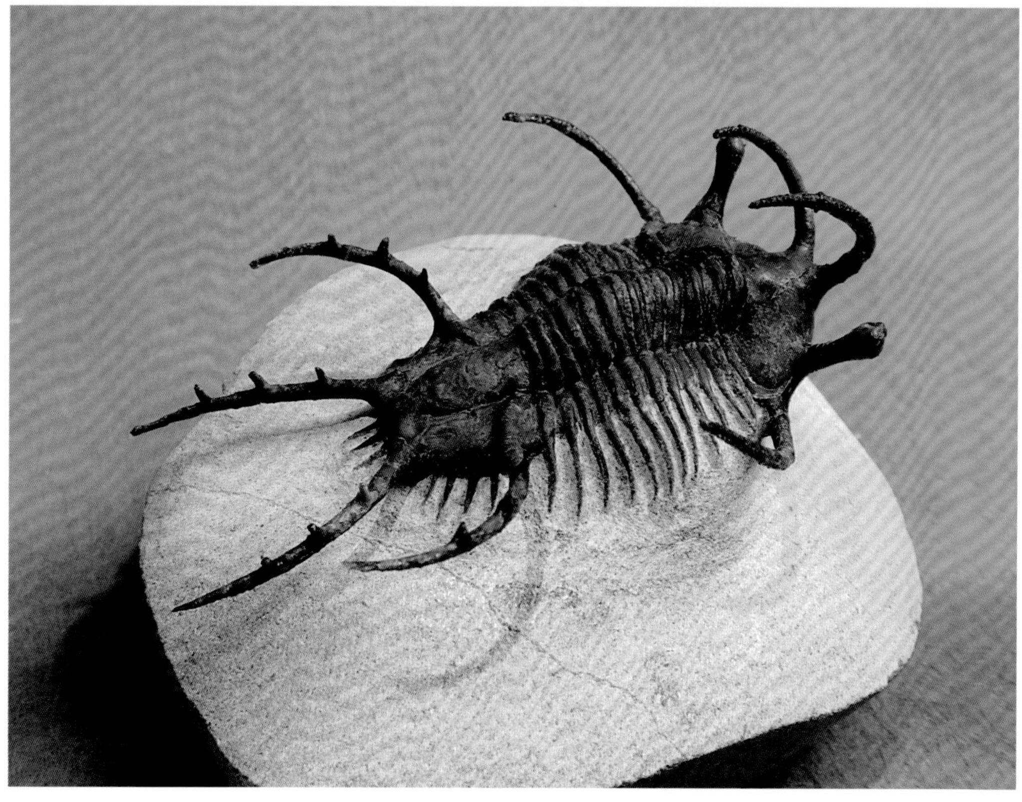

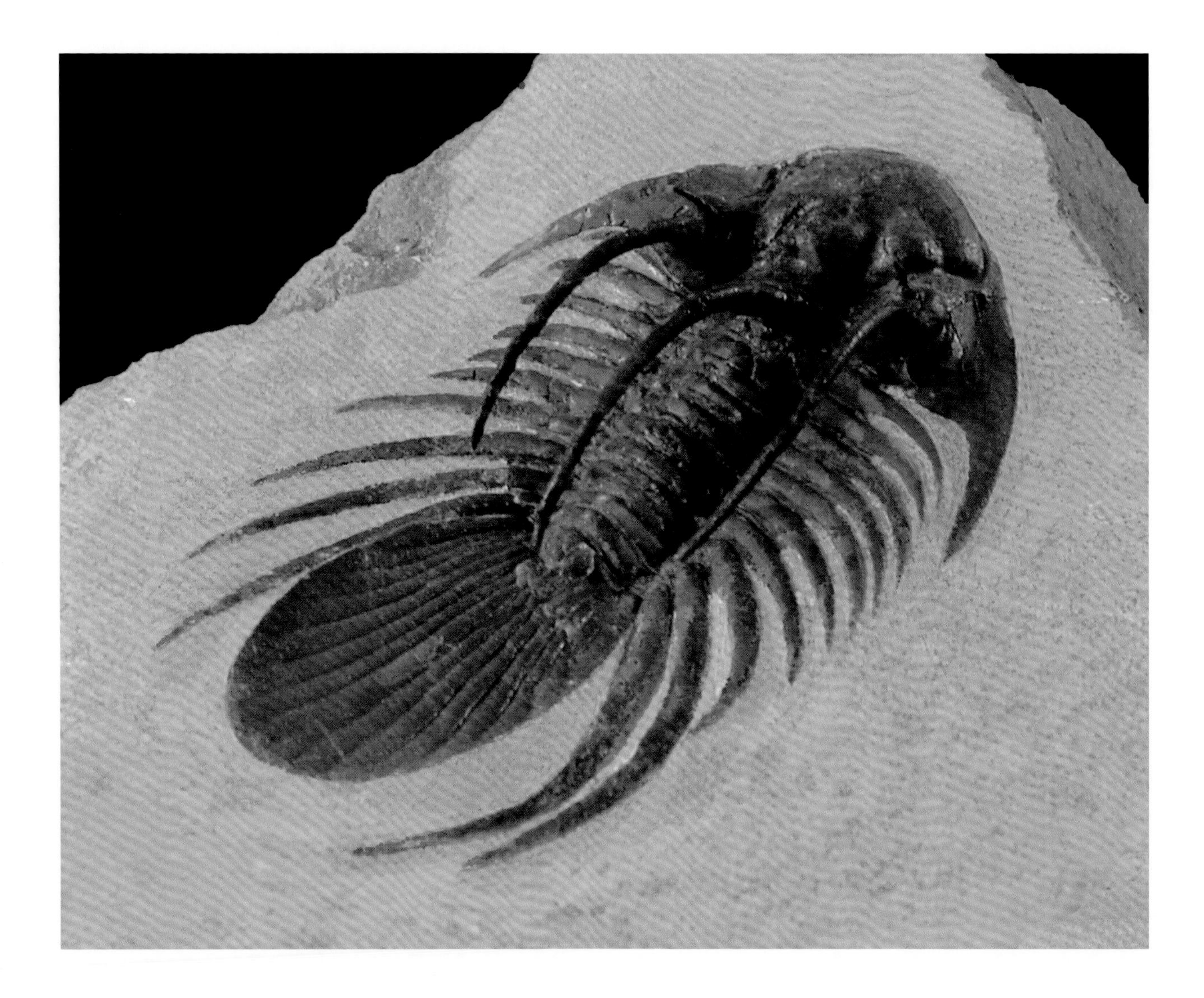

PLATE 90. *Kolyhapeltis hamlaghdadicus* Alberti (36 mm). L. Devonian (Pragian), Hamar Laghdad Limestone, Hamar Laghdad near Erfoud, Morocco. Specimen courtesy of Adam A. Aaronson, photographed by the author.

PLATE 88. *Ceratarges ziregensis* Van Viersen & Presher (32 mm). M. Devonian (Eifelian), El Otfal Formation, Jbel Zireg, Ma'der, south of Jbel Issoumour, Morocco. RLS coll. An accompanying specimen (*upper right*) of *Thysanopeltis* is also exposed from the same matrix.

PLATE 89. *Ceratarges* sp. (42 mm). This variant of the preceding trilobite species, presumably of the same age and also originating from the same general area in the Ma'der, shows short spines protruding from the four major pygidial spines (spines on spines). RLS coll.

PLATE 91. *Ceratonurus* sp. (39 mm, partially enrolled). L. Devonian (Pragian), Hamar Laghdad Limestone, Hamar Laghdad near Erfoud, Morocco. Specimen courtesy of Adam A. Aaronson, photographed by the author. This Odontopleurid trilobite is poorly known; some of its characteristics, in particular the subquadrate cephalon with slender spines on lateral cephalic border, are found in *Ceratocephalus verneuilli* Barrande of the Czech Republic. Surprisingly, this Moroccan species seems to mimic closely the analogous species of *Ceratonurus* of the Haragan Formation of Black Cat Mountain, Oklahoma (Hansen 2009), excavated and prepared with extreme expertise by Bob Carroll.

WESTERN NORTH AMERICA 2.3

I started collecting trilobites over 40 years ago, while venturing into a quarry on the side of the road near McCook, on the outskirts of Chicago, where I stopped to gather some cattails. While roaming through scattered boulders and broken limestone mounds, my eyes stared at a rock bearing a beautiful external impression of a *Calymene niagarensis*, encrusted with sparkling dolomite crystals. This finding rekindled my youthful passion for fossils, which eventually led to the first and second editions of my trilobite books. Alas, no particular color distinguished this or many other trilobites found in these Silurian sediments, where the original cuticle was completely replaced by the same minerals making up the colorless limestone or dolomite matrix. Not much would be added, for the present purposes, by dwelling once more on the images of these trilobites. Instead, the present quest for color evokes other episodes of my search, which took me to many other well-known North American localities, with a preference for the Cambrian exposures of the wild West.

My education about desert environments, not quite on the scale of the Sahara, started with the Mojave Desert of Southern California, in search of the Olenelloid trilobites of the Marble Mountains. Before the advent of four-wheel-drive SUVs, driving through rocky or sandy terrains with a standard rental car proved adventurous. With my friend Mark Utlaut, and the unending patience of my wife Nika, we drove many times from Los Angeles, and several times from Needles at the California-Arizona border, to reach Amboy. This truck stop, with a diner and motel where we lodged (which had an old phone that still had the original one-digit phone number instead of the 10-digit number commonly used today!), was allegedly a favorite of Ronald Reagan, as evident from treasured mementos. The local atmosphere was immortalized at nearby Bagdad, California, in a movie called *Bagdad Café,* starring Jack Palance, and by the view of "fire in the sky" from the solar power station at distant Barstow. From there, we drove to the Marble Mountains through Cadiz, where interminably long Southern Pacific cargo trains stationed, and followed a sandy road along telephone poles to either of two access paths up the mountain to the exposures of the Latham Shale. This greenish shale yielded treasured specimens of the Lower Cambrian *Olenellus fremonti* and other Olenelloid species, illustrated in my previous editions and now revealed in their pristine color. These were the oldest trilobites known, butterflies of the seas, with elegant morphology. My generosity toward a young lady, who was with a group of college students who swamped our diggings, was unfortunately rewarded by her destroying a prized, rare, complete large *Olenellus* after I offered for her to dig at a spot I had just cleared (no good deed shall go unpunished!). In spite of my search efforts, the only images of complete examples of two rare species shown here were kindly provided to me by the courtesy of Norman Brown, a very proficient collector from San Diego. My contact with Norman Brown brought me to another phase of my search for my favorite Lower Cambrian trilobites.

Norman Brown introduced himself to me with a gift by US mail of a delightful specimen of *Olenellus* that he had found at a new Nevada locality and an offer to bring me there in a joint excursion. It turned out that this locality, named Ruin Wash in the Chief Range, had been recently described, with its exceptionally rich Olenelloid fauna, in a paper by Allison Palmer (Palmer 1998). Six different species of these trilobites were described in this paper, representing their terminal appearance in the Lower Cambrian. In-depth studies of this fauna were later carried out by Mark Webster, now a colleague of mine at the University of Chicago. With overwhelming curiosity, I accepted Norman Brown's invitation, and, with my ever-willing trilobite-hunting companion Mark Utlaut, we ventured north of Las Vegas to set base at Caliente in an early, snowy spring. An attraction there was the railroad yard, with its architecturally imposing railway station. From there, we drove toward

Panaca, where a side dirt road took us to the Chief Range where we reached Ruin Wash. This drive in the then greening high desert, toward Bennett Pass, was punctuated by several encounters with families of wild horses staring at us with curiosity. Finally on location, we started rummaging through mounds of shale slabs, which surrounded trenches previously excavated by other visitors. The weathering of exposed shale made it easy to split slabs that resisted previous attempts, and we soon started to be struck with lucky finds. In some of these, a peculiar diagenesis yielded medallions containing a trilobite exoskeleton embedded in a wafer-thin calcite shell, such that the darker enclosed cuticle could still be seen through. When back home, I experimented with applying diluted acids to the surface of the calcite. This progressively dissolved to reveal in stunning contrast the enclosed almost black trilobite cuticle. Such treated medallions yielded pictorial, colorful images, in particular when immersed in xylene, as shown here. In spite of several further returns to this site, with my sons Emile and Matteo, the best examples of these trilobites shown here were borrowed from the collection of Norman Brown, who kindly let me exert my photographic preferences. Other unparalleled specimens originate from the Tucson Gem and Mineral Show, where no trilobite ever escapes being found. Another locality close to Ruin Wash where we spent some time was Klondike Gap. The exposed layers, older than those at Ruin Wash, belong to the Delamar Member of the Pioche Formation and contain a trilobite fauna that is identical to that of the Latham Shale found several hundred miles west, in the Marble Mountains of San Bernardino County, California

A majestic environment of Upper Middle Cambrian depositional layers in the House Range of western Utah is the Wheeler Amphitheater, near Antelope Springs. This has been a mecca for trilobite collectors over many decades, with its exposures of the Wheeler Formation in its central part and surrounding peaks of the Marjum Formation overlaying the Weeks Formation at Marjum Pass. Robert Harris at Delta, Utah, opened a commercial quarry in the Wheeler Shale in the early 1960s, which yielded millions of the famous *Elrathia kingii* trilobites, sold mainly at his Bug House in Delta and in his exhibits at the Tucson Show and many others. These are the "padded" trilobites with the peculiar "cone in cone" calcite diagenesis, mentioned in my introduction. Robert Harris was kind enough to guide me not only to his quarries but also to several other surrounding localities of the Marjum Formation, which I visited with many friends over the years. Specimen of *Elrathia*, *Asaphiscus*, and the Agnostid *Peronopsis* could be easily found not only in Harris's quarries but also in matrix—free, lying on the terrain surface almost anywhere nearby. Other exposures in the Wheeler Amphitheater yielded exclusively *Bathyuriscus* and *Modocia*. Weather adversities were pretty much the norm but did not deter my searching. More serious was once the loss of brake hydraulics on our rental car, due to a protruding rock on the dirt road, while climbing to Marjum Pass with my friend Mark Utlaut. Somehow we made it to Marjum Pass, where we left our damaged car and started climbing a steep ravine toward a quarry high above. The climb was embellished by the dried-up, twisted roots and trunks of ancient juniper trees (*Juniperus osteosperma*), with their contorted branches still pointing to the sky, a memorable sight. The quarry looked impervious, with walls and floor of hard limestone. We managed to lift some blocks with a crowbar inserted in providential cracks, and finally, using a sledge-hammer, we cracked some blocks open, with exultant exclamations. They contained complete specimens of *Hemirodon amplipyge*, surrounded by colorful mineralizations, a prized find. On another trip to Antelope Springs with a young couple—a former student of mine, Shelly St. Louis-Weber (now a most successful manager at Intel), and husband Brant—I made an unforgivable mistake on the highway from Delta toward an exit that should have taken us to our destination. I exited on a side road too soon and ended up trapped in the mud surrounding a pond, in spite of our driving an SUV. We were assaulted by swarms of avid mosquitoes that endangered our lives while walking back to the main highway to hitchhike a ride back to Delta. A couple of Good Samaritans drove us back, two at a time, to Delta where, amidst local amusement, we had to hire a truck to get our car out of the sinking trap. Unexpected to get stuck in mud in a desert!

Besides my excursions into the Cambrian of the wild West, there have been a large number of hunting trips into US trilobite localities belonging to subsequent geological periods. Within driving distances from my home base in Chicago have been Ordovician, Silurian, and Devonian localities of the US Midwest and adjacent Ontario. Much time was spent in stone quarries, not particularly picturesque under relentless sun and no shade, often accompanied by my sons Emile and Matteo, who were of great help in spotting trilobites lurking in the rocks. Many trips took us to Ordovician localities in southern Indiana and Waynesville and Westwood in Ohio in search of *Flexicalymene* and *Isotelus*. On the shore of Lake Huron, in Collingwood, Ontario, we found large slabs of black slate filled with the remains of *Pseudogygites latimarginatus*, which weather to bright light colors. Aside from the local Silurian limestone quarries in the vicinity of Chicago, more colorful Silurian trilobites were extracted in a large quarry at Waldron, Indiana, where, in addition to *Calymene*, attractive *Dalmanites* exuviae were a frequent find.

Most rewarding have been many of our trips to the Devonian quarries at Sylvania, Ohio, and further east, to the so-called Grabau beds, on Lake Erie, south of Buffalo, New York. The quarries of the Medusa Cement Quarry at Sylvania, Ohio, are famous for the spectacular preservation of the Phacopid trilobites found in the layers of Silica Shale, which occur interleaved with thick layers of limestone. The soft shale is discarded in giant heaps where it weathers to mud, and abundant pyritized brachiopods and especially enrolled trilobites are released intact. These often collect in rivulets, just ripe for picking. In blocks of freshly mined shale, outstretched complete trilobites, sometimes in large assemblies, can be extracted. These are mostly *Phacops rana* (the original taxonomy), which occur in two varieties called *milleri* and *crassituberculata*, differing in the number of schizochroal eye lenses (a special section of this book is devoted to the illustration of trilobite eyes). The genus name has recently been replaced with *Eldredgeops*, to honor the paleontologist Niles Eldredge, who in his doctoral thesis provided an authoritative, exhaustive study of these trilobites. Fascinating to the collector is the dark green, translucent appearance of their carapace, sometimes partly pyritized in golden hues, starkly contrasted against the light gray of the encasing matrix. The best collecting occurred just after major mining, which usually occurred on Friday afternoon, leaving large areas of scattered boulders for swarms of collectors on Saturday mornings. It took a keen eye to discern the periodic pattern of the trilobite tergites, covered by a layer of white dried mud, occasionally exposed on the surface of large boulders. I cannot forget a perfect finding of mine (plate 196 of my second edition), spotted in such form on top of a large block of shale while I was trying to find an opening among a multitude of crouched, hammering collectors. This made my day and was the cause of much envy on part of the surrounding crowd.

A much more scenic occurrence of similar Hamilton Group trilobite fauna is found in the Grabau beds, on the coast of Lake Erie, a few miles south of Buffalo, New York. To reach the locality, on the grounds of a Piarist seminary, located in a Frank Lloyd Wright villa (Graycliff) close to the edge of a cliff overlooking the lake, permission was granted by the sympathetic director of the seminary. A rickety, winding iron staircase on the edge of the cliff, some 100 steps, was hesitantly climbed to reach a narrow beach, lapped by gentle waves. Fragments of shale fallen from the vertical cliff wall had accumulated on the inner edge of the beach, soon found to contain smallish *Phacops* fragments. More rewarding was to watch the large shelf of shale gently protruding into the lake, periodically exposed under shallow waves, to spot these black trilobites, still encased in the rock, emerging here and there underwater. It was possible to chisel out tablets of shale carrying the exposed trilobite (see, e.g., plate 143) by wading barefoot on the shelf. Among these, occasional *Greenops,* with their starlike pygidium, were also captured. This was the most unusual trilobite hunting I was ever exposed to and is still cherished in my memory.

Among other unforgettable memories was a visit to Black Cat Mountain in the Devonian Haragan Formation of Oklahoma, which was literally carved out of the wilderness by my friend Robert (Bob) Carroll. He is a well-known, devoted trilobite enthusiast and dealer, whom I met many years ago at the Tucson Show and who credited my trilobite books for having unveiled his true vocation. Bob is a most skillful artist in the preparation of delicate three-dimensional structures of beautiful *Kettneraspis* (formerly *Leonaspis*), *Ceratonurus*, *Dicranurus*, and many other local trilobites, carved out of a hard limestone with vibrotools and miniature pneumatic airbrasive tools. Several images of his prepared trilobites enriched the pages of the second edition of my trilobite book, together with those of astonishing spiny Moroccan specimens of *Psychopyge* and *Philonyx*. A couple of years ago, with my wife Nika and our friend Scott Evans, we decided to pay him a visit at his home base in Clarita, Oklahoma. He took us to his quarry, winding up to the top of Black Cat Mountain (named after a beloved pet buried there), and we started looking for trilobites in the heavy slabs of limestone that he had mined out. It soon became obvious that finding any exposed specimen was, for us, an impossible task. We marveled at Bob's ability to fracture this hard rock and to spot evidence of any encased specimen, desperately hidden from our unacquainted sight. It was a revelation to watch him, back in his prep lab, slowly unveil the existence of a beautiful trilobite on a block of rock, with his airbrasive tools. We did learn from him the use of miniature sand blasters and the art of making bubble-free latex molds to obtain perfect casts of some of his treasures.

Finally, the Tucson Show provided me with colorful images of North American trilobites that originated from several other notable locations that were visited without much success or that were beyond my reach. For these contributions, I am grateful to many friends whom I met over the years.

CAMBRIAN

PLATE 92. *Olenellus (Mesonacis) fremonti* Walcott (82 mm). L. Cambrian, Latham Shale of the Marble Mountains, San Bernardino County, California. RLS coll. This color image completes my previous black-and-white description of this specimen (Levi-Setti 1993, plate 54). Following Lieberman's *Systematic Revision of the Olenelloidea* (Lieberman 1999), this species is definitely assigned to the genus *Mesonacis*.

PLATE 93. *Olenellus (Mesonacis) fremonti* Walcott (55 mm). L. Cambrian, Latham Shale of the Marble Mountains, San Bernardino County, California. RLS coll. I always cherished this attractive specimen, found staring at me while climbing a heap of debris.

PLATE 94. *Olenellus (Mesonacis) fremonti* Walcott (width of largest cranidium 105 mm). L. Cambrian, Latham Shale of the Marble Mountains, San Bernardino County, California. RLS coll. These complete cranidia portray the anterior cranidial border, missing in the specimens figured in the preceding two plates.

PLATE 95. *Olenellus clarki* (Resser) (28 mm). L. Cambrian, Latham Shale of the Marble Mountains, San Bernardino County, California. RLS coll. Partially disarticulated exuviae, previously shown in their black-and-white version (Levi-Setti 1993, plate 56).

PLATE 96. *Olenellus clarki* (Resser) (largest individual 17 mm long). L. Cambrian, Latham Shale of the Marble Mountains, San Bernardino County, California. RLS coll. An assembly of juvenile complete individuals.

PLATE 97. *Bristolia bristolensis* (Resser) (36 mm). L. Cambrian, Delamar Member of the Pioche Formation, exposed at Klondike Gap in the Chief Range of Lincoln County, Nevada. The trilobite fauna of this outcrop is identical to that of the age equivalent Latham Shale of the Marble Mountains of San Bernardino County, California. To be noted here is the subquadrate profile of the posterior cranidial border (see also Levi-Setti 1993, plates 60 and 61c) seen in the complete individual and in two separated cranidia in the same image. This subquadrate profile differs significantly from that (V-shaped) attributed to this species of *Bristolia* by Palmer and Repina (Palmer and Repina 1993, fig.4.5) and also reproduced in R. C. Moore's *Treatise on Invertebrate Paleontology,* Part O, Revised (1997), fig. 258.1. Photograph courtesy of Norman Brown, Norman Brown coll.

PLATE 98. *Bristolia mohavensis* (Hazzard & Crickmay) (20 mm). L. Cambrian, Latham Shale of the Marble Mountains, San Bernardino County, California. A cast of this specimen was recently shown by Webster (Webster 2011, fig. 9.1). A morphologically similar example of this trilobite was previously attributed to *Olenellus mohavensis* (Resser) (see, e.g., Levi-Setti 1993, plate 59). Also closely similar to *Bristolia mohavensis* is the specimen figured as *Bristolia bristolensis* (Resser), as referred to in the caption to the preceding plate. Specimen courtesy of Norman Brown, photographed by the author. Norman Brown coll.

PLATE 99. *Bristolia insolens* Resser (34 mm). L. Cambrian, Latham Shale of the Marble Mountains, San Bernardino County, California. Specimen courtesy of Norman Brown, photographed by the author.

PLATE 100. *Olenellus (Olenellus) gilberti* Meek (37 mm). L. Cambrian, Pioche Formation, Ruin Wash location in the Chief Range, Lincoln County, eastern Nevada. This richly fossiliferous locality represents the "Terminal Early Cambrian Extinction of the Olenellina" (Palmer 1998). Specimen courtesy of Norman Brown, photographed by the author. The carapace of this trilobite is encased in a thin calcitic medallion, slightly eroded in diluted formic acid to better expose its dark color. This image was taken in air, to be compared with the next plate, which was photographed while the sample was immersed in xylene.

PLATE 101. *Olenellus (Olenellus) gilberti* Meek. This image portrays the same trilobite shown in the preceding plate, photographed while immersed in xylene. This technique, used in several examples illustrated in the second edition of my trilobite book (Levi-Setti 1993), takes advantage of the elimination of light reflections at the air-object interface, establishing optical contact that can reveal subsurface details of a semitransparent medium. In this case, the interior of the prothoracic axis is seen to contain a tubular structure (possibly alimentary canal) otherwise unseen in the preceding image.

PLATE 102. *Olenellus (Olenellus) gilberti* Meek (48 mm). L. Cambrian, Pioche Formation, Ruin Wash location in the Chief Range, Lincoln County, eastern Nevada. RLS coll. Partially disarticulated individual, internal mold, devoid of carapace remains.

PLATE 103. *Olenellus (Paedumias) chiefensis* Palmer (42 mm). L. Cambrian, Pioche Formation, Ruin Wash location in the Chief Range, Lincoln County, eastern Nevada. RLS coll. A teratology is present on the left pleura of the third thoracic segment.

PLATE 104. *Olenellus (Paedumias) chiefensis* Palmer (29 mm). L. Cambrian, Pioche Formation, Ruin Wash location in the Chief Range, Lincoln County, eastern Nevada. Specimen courtesy of Mark Webster, photographed by the author. The posterior-most section of the prothorax is folded over, possibly following partial enrollment (also shown in fig. 6A of Webster 2008).

PLATE 105. *Olenellus fowleri* Palmer (65 mm). L. Cambrian, Pioche Formation, Ruin Wash location in the Chief Range, Lincoln County, eastern Nevada. RLS coll. External impression, exhibiting fine details of the prosopon (thin ridges radiating from the eye lobes into the genal fields of the cephalon, also called genal caeca), as well as a long, multisegmented opisthothorax.

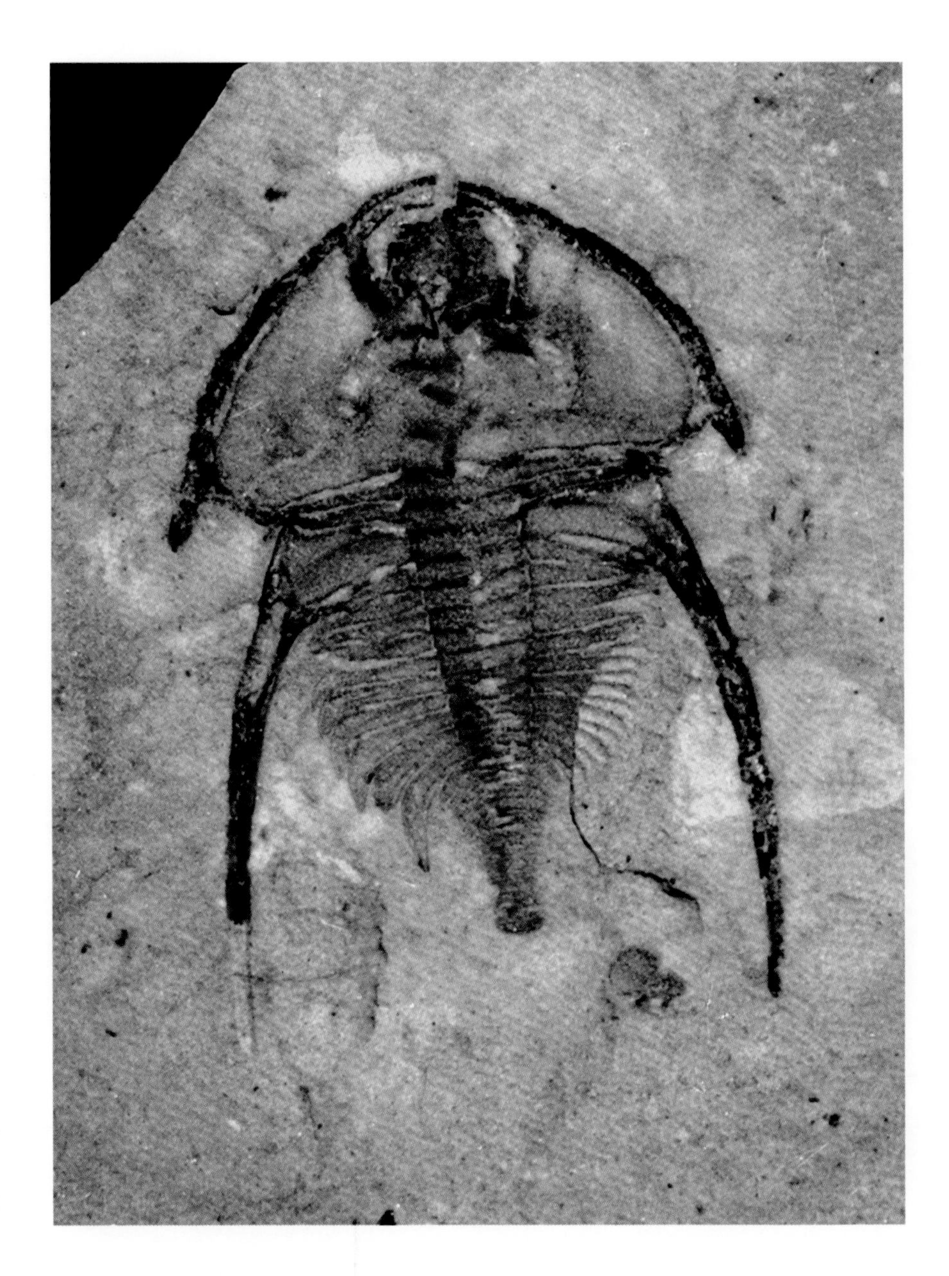

PLATE 107. *Nephrolenellus geniculatus* Palmer. The same specimen as the preceding plate, photographed while immersed in xylene. Mineralized regions of the carapace appear in stark contrast by this method.

PLATE 106. *Nephrolenellus geniculatus* Palmer (21 mm). L. Cambrian, Pioche Formation, Ruin Wash location in the Chief Range, Lincoln County, eastern Nevada. RLS coll. Specimen photographed in air. Opisthothorax present, seemingly abruptly terminated.

PLATE 108. *Olenellus roddyi* Resser & Howell (51 mm). L. Cambrian, Forteau Formation, near Roddinckton, western Newfoundland. This species, known from the Kinzers Formation of Pennsylvania, seems close to *Olenellus clarki* (Resser). RLS coll. In this plate, the specimen was photographed in air, so that the image can be compared with that in the next plate.

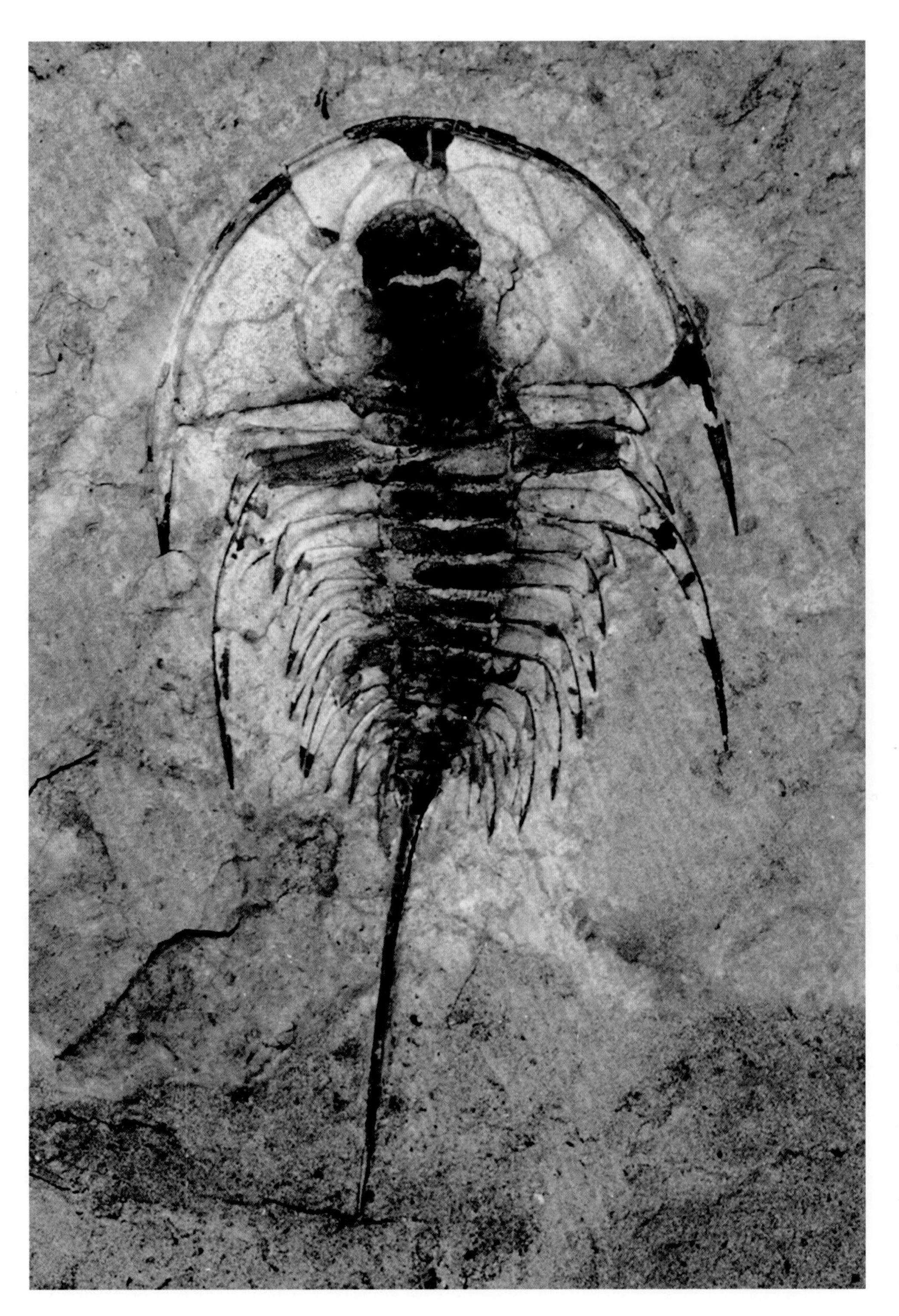

PLATE 109. *Olenellus roddyi* Resser & Howell. The same specimen of the preceding plate, photographed while immersed in xylene. In addition to revealing details of mineralization not highlighted in the preceding plate, this image clearly shows that the cephalon is displaced forward and off axis. This suggests an early stage of ecdysis or exuviation.

PLATE 110. *Nevadia weeksi* Walcott (53 mm). L. Cambrian, Montenegro Member of the Campito Formation in Esmeralda County, Nevada. Specimen courtesy of Norman Brown, photographed by the author.

PLATE 111. *Wanneria walcottana* (Wanner) (49 mm). L. Cambrian, Kinzers Formation, Lancaster, Pennsylvania. Specimen collected by Michael Thomas of York, Pennsylvania. RLS coll. External impression, not showing surface ornamentation. The counterpart of this specimen (internal impression) was shown previously in a black-and-white image (Levi-Setti 1993, plate 63).

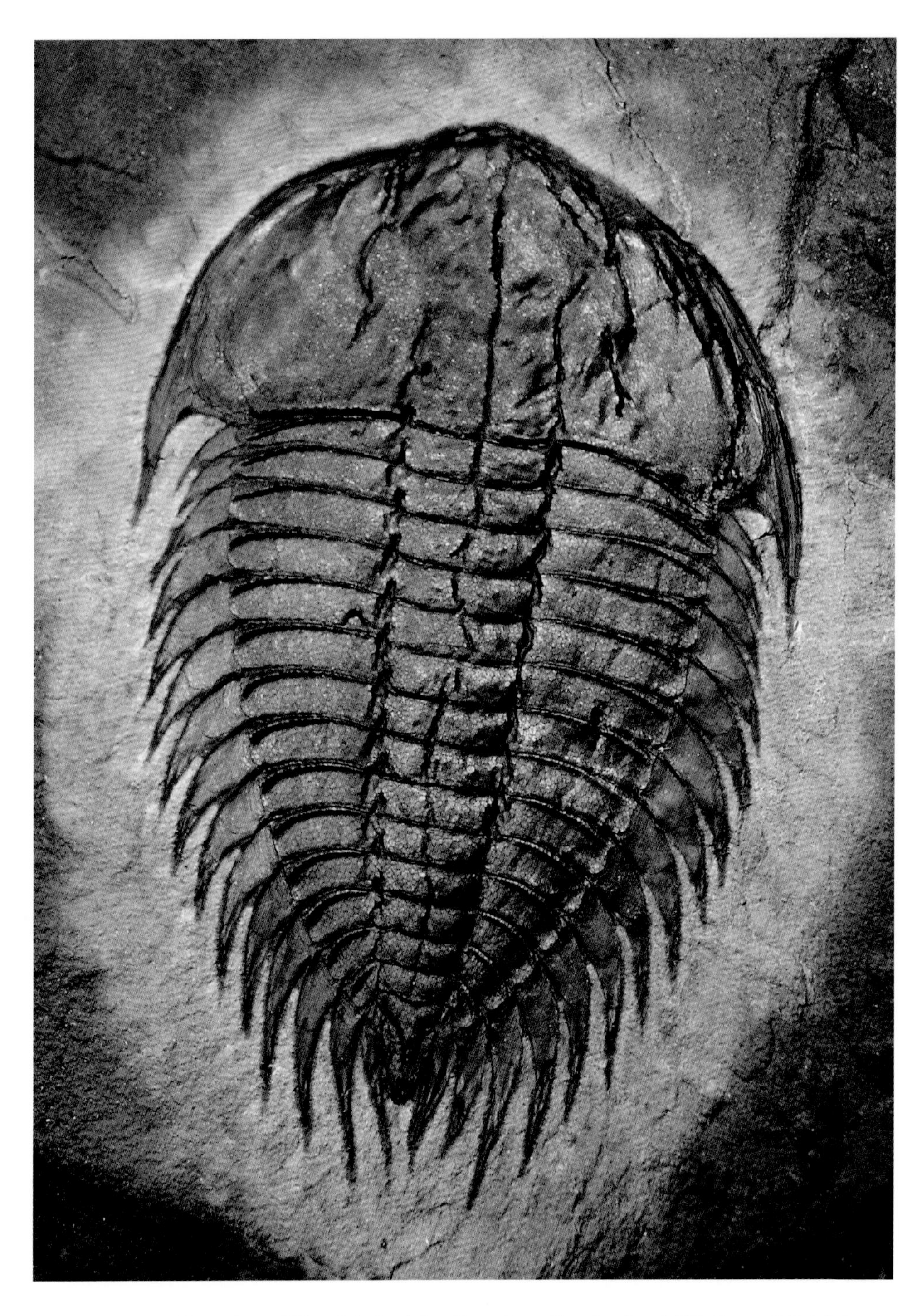

PLATE 112. *Wanneria* sp. (122 mm). L. Cambrian, Eager Formation, Cranbrook Rifle Range, British Columbia, Canada. This beautiful specimen made a brief appearance at the Tucson Gem and Mineral Showin 2003, where I had a chance to take this photograph. The species is referred to as *Wanneria dunnae* Bohach at several entries on the web, from the unpublished PhD thesis by Lisa Bohach.

PLATE 113. *Zacanthoides typicalis* Walcott (larger individual 34 mm). M. Cambrian, Chisholm Shale, Half Moon Mine, Mount Ely, Pioche, Nevada. Photograph courtesy of Norman Brown. Norman Brown coll. The notable asymmetry of the right-hand side of the cranidium, in the larger specimen, may represent a teratology or may be due to the overlapping of a shifted librigena. The smaller individual is missing the posterior part of the thorax beyond the eighth axial ring, which carries the long axial spine.

PLATE 114. *Zacanthoides typicalis* Walcott. M. Cambrian, Chisholm Shale, Half Moon Mine, Mount Ely, Pioche, Nevada. Photograph courtesy of Norman Brown. Norman Brown coll. An unusual side view of this trilobite, illustrating the posture of the axial spine.

PLATE 115. *Elrathia kingii* (Meek) (55 mm). M. Cambrian, Wheeler Shale, Wheeler Amphitheater, Antelope Springs, Millard County, Utah. RLS coll. This is the largest individual in my collection. Visible on the cephalon are the fine ridges of the genal caeca (see caption to plate 105). Although the exoskeleton of these trilobites may be only a fraction of a millimeter thick, in the trilobites from this location a particular form of diagenesis is taking place, making the carapace appear much thicker. Acicular calcite crystals grow in concert beneath the cuticle, projecting faithfully its internal surface by several millimeters (cone-in-cone structure) (Bright 1959).

PLATE 116. *Elrathia kingii* (Meek) (largest 28 mm). M. Cambrian, Wheeler Shale, Wheeler Amphitheater, Antelope Springs, Millard County, Utah. RLS coll. Slab showing various stages of growth of this trilobite.

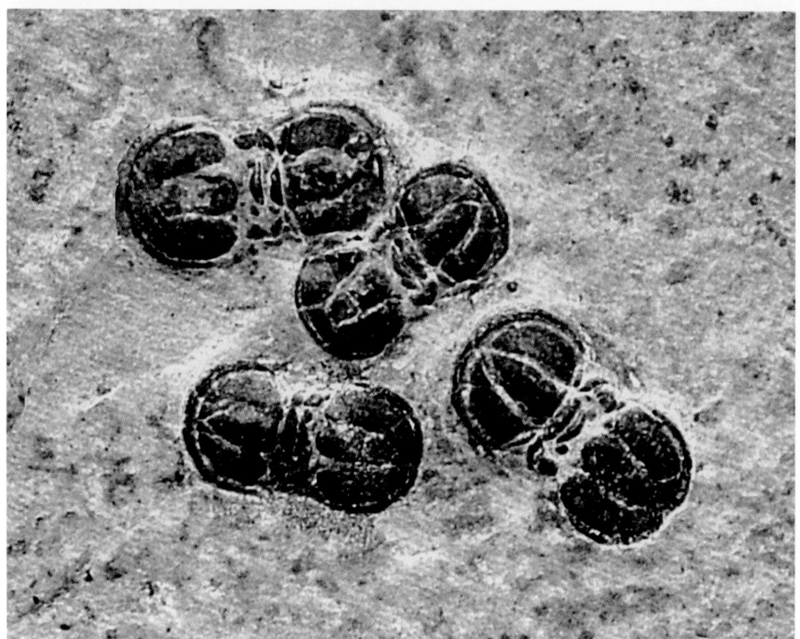

PLATE 117. *Elrathia kingii* (Meek) (large individuals 9 mm). M. Cambrian, Wheeler Shale, Wheeler Amphitheater, Antelope Springs, Millard County, Utah. RLS coll. A cluster of small individuals, some a few millimeters long.

PLATE 118. *Peronopsis interstricta* (White) (each 6 mm). M. Cambrian, Wheeler Shale, Wheeler Amphitheater, Antelope Springs, Millard County, Utah. RLS coll. This diminutive agnostid trilobite is often found in association with the population of *Elrathia kingii.*

PLATE 119. *Asaphiscus wheeleri* Meek (48 mm). M. Cambrian, Wheeler Shale, Wheeler Amphitheater, Antelope Springs, Millard County, Utah. RLS coll. This companion of the *Elrathia kingii* beds came as a calcite cone-in-cone concretion that neatly split to reveal the two sides of the uncased trilobite exoskeleton.

PLATE 120. *Alokistocare harrisi* Robison (58 mm). M. Cambrian, Wheeler Shale, Wheeler Amphitheater, Antelope Springs, Millard County, Utah. Specimen courtesy of Jason Cooper, photographed by the author at the 2009 Tucson Gem and Mineral Show. The carapace surface of this perfect specimen is covered by small granules.

PLATE 122. *Modocia typicalis* (Resser) (44 mm). M. Cambrian, Marjum Formation, locality close to that of the preceding plate in the House Range, Utah. RLS coll. Specimen collected by Robert Harris, previously seen in a black-and-white photograph (Levi-Setti 1993, plate 107). Isolated complete specimens of this trilobite are found in a quarry of hard, large tabular shale slabs of tan color.

PLATE 121. *Bathyuriscus fimbriatus* Robison (58 mm). M. Cambrian, Lower Marjum Formation, House Range next to the Wheeler Amphitheater, Utah. RLS coll. Copious examples of this trilobite are found in layers located a few kilometers southwest of the Elrathia quarries, next to the road leading to Marjum Pass. This specimen is missing the librigena, cast in molting.

PLATE 123. *Hemirodon amplipyge* Robison (86 mm). M. Cambrian, Marjum Formation, Marjum Pass, House Range, Utah. RLS coll. Specimen collected by Robert Harris. My adventures to reach the Hemirodon quarry at Marjum Pass are discussed in the introduction to section 2.3. Although several complete specimens of this trilobite were retrieved and stored in my collections, none had the appeal of the sample chosen for inclusion in this context.

PLATE 124. *Hemirodon amplipyge* Robison (75 mm). M. Cambrian, Marjum Formation, Marjum Pass, House Range, Utah. Specimen courtesy of Dave Douglass, photographed by the author. Once again, I was captivated, in this case by the aesthetic, colorful appeal of the matrix surrounding the enclosed trilobite.

PLATE 126. *Olenoides superbus* (Walcott) (specimen at left 106 mm). M. Cambrian, Marjum Formation, south edge of Wheeler Amphitheater, House Range, Utah. Specimen courtesy of Jake Skabelund, photographed by the author at the 2012 Tucson Gem and Mineral Show.

PLATE 125. *Olenoides inflatus* Walcott (116 mm). M. Cambrian, Marjum Formation, south edge of Wheeler Amphitheater, House Range, Utah. Specimen courtesy of Jason Cooper, photographed by the author at the 2009 Tucson Gem and Mineral Show.

PLATE 127. *Tricrepicephalus texanus* (Shumard), formerly *Tricrepicephalus coria* (Walcott) (39 mm). U. Cambrian, Weeks Formation, House Range, Millard County, Utah. Specimen courtesy of Uberle, photographed by the author at the 2011 Tucson Gem and Mineral Show.

PLATE 128. *Weeksina unispina* (Walcott) (14 mm). U. Cambrian, Weeks Formation, House Range, Millard County, Utah. Specimen courtesy of Dave Douglass, photographed by the author.

ORDOVICIAN

PLATE 129. *Homotelus bromidensis* Esker (41–49 mm). M. Ordovician, Bromide Formation (Poolville Member), Blackriverian, Critter Hill, Carter County, Oklahoma. RLS coll. A densely packed assembly.

PLATE 130. *Ceraurus pleurexanthemus* Green (specimen at upper left 37 mm). M. Ordovician, Bobkaygeon, Carden, Ontario, Canada. Specimen courtesy of Dave Douglass, photographed by the author. A detail of this slab (left third of slab) was previously shown in a black-and-white image (Levi-Setti 1993, plate 149). Several associated specimens of *Flexicalymene* are also present in this slab.

PLATE 131. *Isotelus gigas* De Kay (79 mm). M. Ordovician, Trenton group, Trenton Falls, New York. RLS coll.

PLATE 132. *Flexicalymene meeki* (Foerste) (34 mm). U. Ordovician, Waynesville Formation (Fort Ancient Member), Westwood, Cincinnati, Ohio. RLS coll. This specimen is seen as it was extracted from a flooded exposure, with no preparation.

PLATE 133. *Flexicalymene meeki* (Foerste) (upper left 19 mm horizontal, lower right 17 mm horizontal). U. Ordovician, Waynesville Formation (Fort Ancient Member), Westwood, Cincinnati, Ohio. RLS coll. Two enrolled specimens, unprepared.

PLATE 134. *Pseudogygites latimarginatus* (Hall) (55 mm). U. Ordovician, Lower Whitby Formation, Bowmanville, Ontario, Canada. RLS coll.

PLATE 136. *Dalmanites limuloides* Green (image 310 ✣ 780 mm^2). Silurian, Rochester Shale Formation, Lockport, New York. Specimen courtesy of Lang's Fossils, photographed by the author at the 2006 Tucson Gem and Mineral Show.

SILURIAN

PLATE 135. *Arctinurus boltoni* (Bigsby) (115 mm). Silurian, Rochester Shale Formation, Middleport, New York. Specimen courtesy of Jake Skabelund, photographed by the author at the 2012 Tucson Gem and Mineral Show.

PLATE 137. *Dalmanites limuloides* Green (each 71 mm). Silurian, Rochester Shale Formation, Lockport, New York. Specimen courtesy of Lang's Fossils, photographed by the author at the 2006 Tucson Gem and Mineral Show. Magnified image of a cluster seen in the preceding plate.

PLATE 138. *Calymene niagarensis* Hall (41 mm). Silurian, Racine Formation in the Niagaran Limestone, McCook, Illinois. RLS coll. Negative mold encrusted with dolomite crystals. In this and many other specimens collected by the author at the above location, a hollow cavity is found in place of the steinkern. Image obtained with illumination from the northwest.

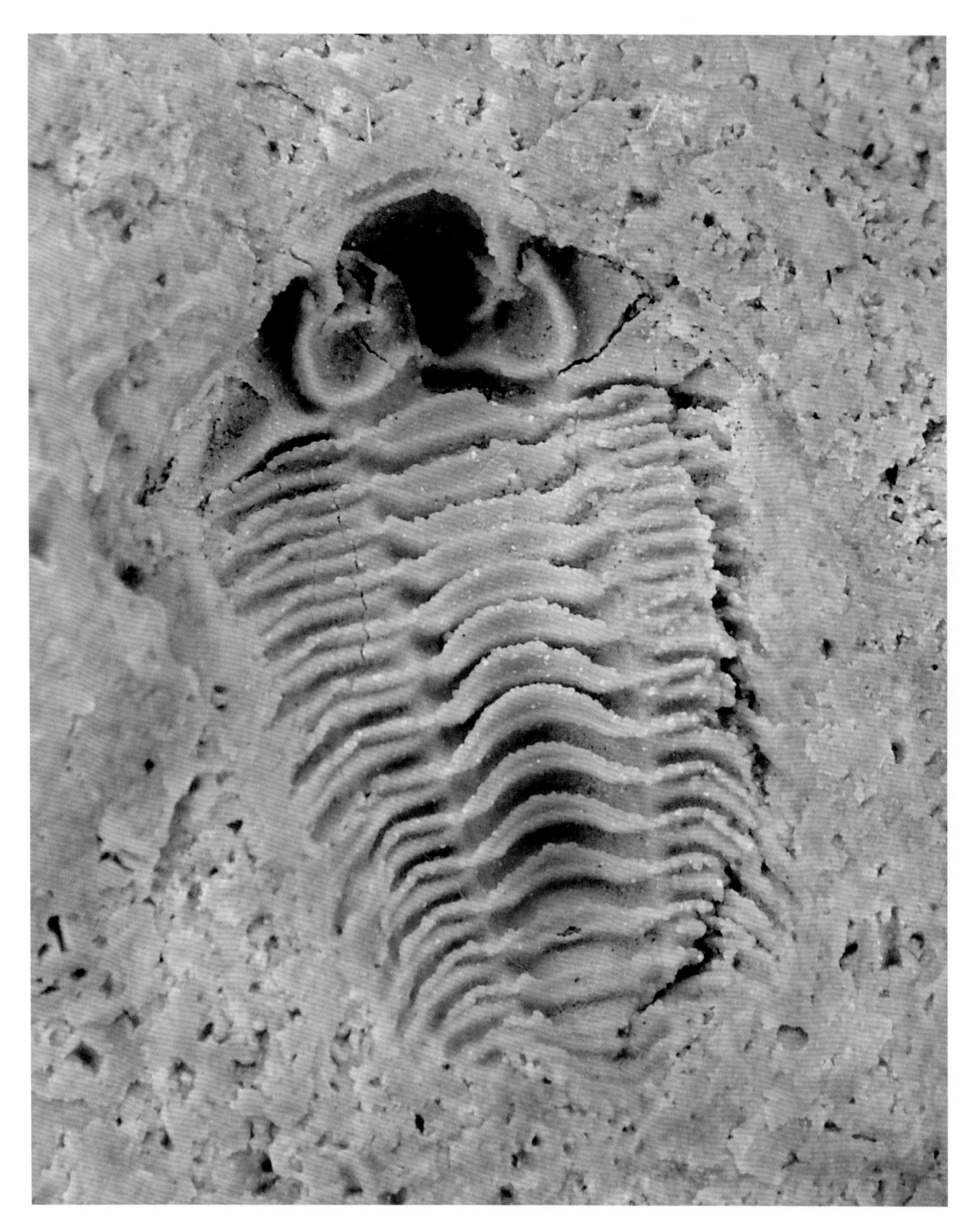

PLATE 139. Same as plate 138, with specimen illuminated from the southeast. With this illumination, depressions transform into relief, like that of a cast. Comparison of the two images reveals details missed in either one.

DEVONIAN

PLATE 140. *Eldredgeops rana milleri* Stewart, formerly *Phacops rana milleri* (65 mm). Devonian (Cazenovian), Silica Shale Formation, Hamilton Group, Medusa Portland Cement Quarry, Silica near Sylvania, Ohio. RLS coll. This is my "golden eye" trilobite (the left eye of this specimen is pyritized). The schizochroal eyes of these beautiful trilobites are discussed in section 3 of this book.

PLATE 141. *Eldredgeops rana milleri* Stewart (26 mm). Same origin as that of the preceding plate, this time enrolled, showing details of the beautifully preserved eye. RLS coll. This posture shielded the specimens from damage in the easily waterlogged shale such that enrolled individuals could be found intact in local water rivulets descending from heaps of quarry dumps.

PLATE 142. *Eldredgeops* in a spectacular assemblage from the Silica Shale of Silica, Ohio. This grouping was previously shown in a black-and-white image (Levi-Setti 1993, plate 199) photographed by the author through courtesy of David C. Rilling. This specimen is now at the Smithsonian Institution, National Museum of Natural History, and the present color image was obtained through courtesy of Thomas Jorstad of its Department of Paleobiology.

PLATE 143. *Phacops rana* Green (26 mm). M. Devonian, Moscow Formation, Hamilton Group, "Grabau beds," shore of Lake Erie, near Buffalo, New York. RLS coll. A colorful description of my adventures at this location is given in the introduction to the present section of this book. This lovely small trilobite was extracted as seen here, without need of any preparation.

PLATE 144. *Phacops rana* Green (24 mm each). M. Devonian, Moscow Formation, Hamilton Group, "Grabau beds," shore of Lake Erie, near Buffalo, New York. RLS coll. Of this graceful trio, the center one is a steinkern, the two at the sides are external impressions.

PLATE 146. *Bellacartwrightia jennyae* (Lieberman & Kloc) (44 mm). M. Devonian, Centerfield Limestone, York, New York. Specimen courtesy of Wayne Ciluffo, photographed by the author at the 2012 Tucson Gem and Mineral Show.

PLATE 145. *Greenops (Greenops) boothi* (Green) (26 mm). M. Devonian, Moscow Formation, Hamilton Group, "Grabau beds," shore of Lake Erie, near Buffalo, New York. RLS coll. A rare occurrence of this elegant trilobite, cleaned up by the gentle waves of Lake Erie. The cephalon of a *Phacops rana* overlaps the left front side.

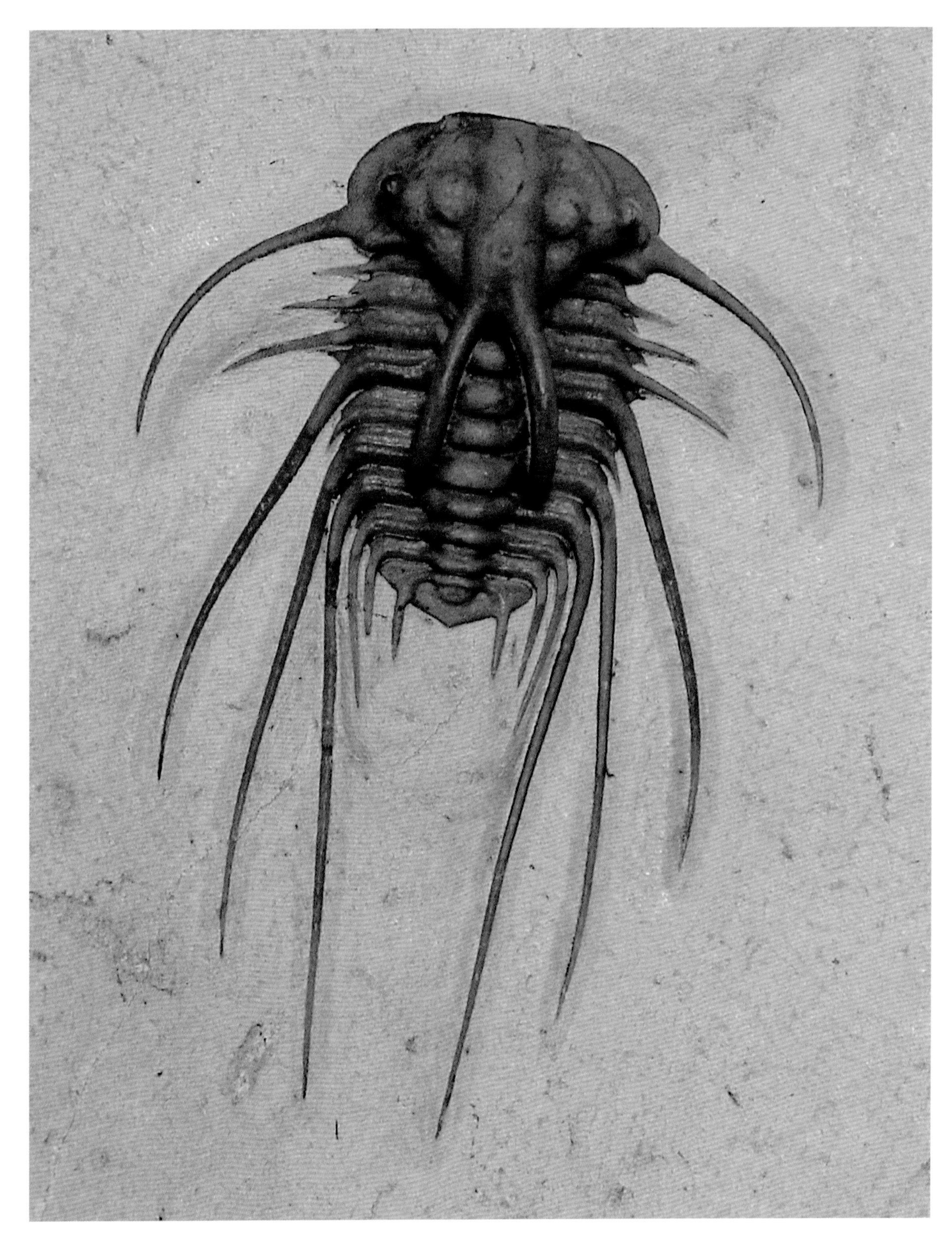

PLATE 147. *Dicranurus hamatus elegantus* (Campbell) (36 mm). L. Devonian, Haragan Formation, Coal County, Black Cat Mountain, Clarita, Oklahoma. Specimen courtesy of Robert (Bob) Carroll, photographed by the author. The legendary expertise of Bob Carroll in the preparation of the complex and delicate architecture of spiny trilobites, extracted from massive hard limestone, is discussed in the introduction to the present section of this book. It has also been expertly illustrated in a comprehensive book dedicated to the trilobite fauna of Black Cat Mountain (Hansen 2009).

PLATE 148. *Kettneraspis williamsi* Whittington, formerly *Leonaspis williamsi* (17 mm). L. Devonian, Haragan Formation, Coal County, Black Cat Mountain, Clarita, Oklahoma. RLS coll. This is another of Bob Carroll's little jewels, beautifully prepared. Only by actually attempting to remove the encasing limestone matrix from a barely surfacing protrusion with an airbrasive tip, can one appreciate the difficulties involved in this process.

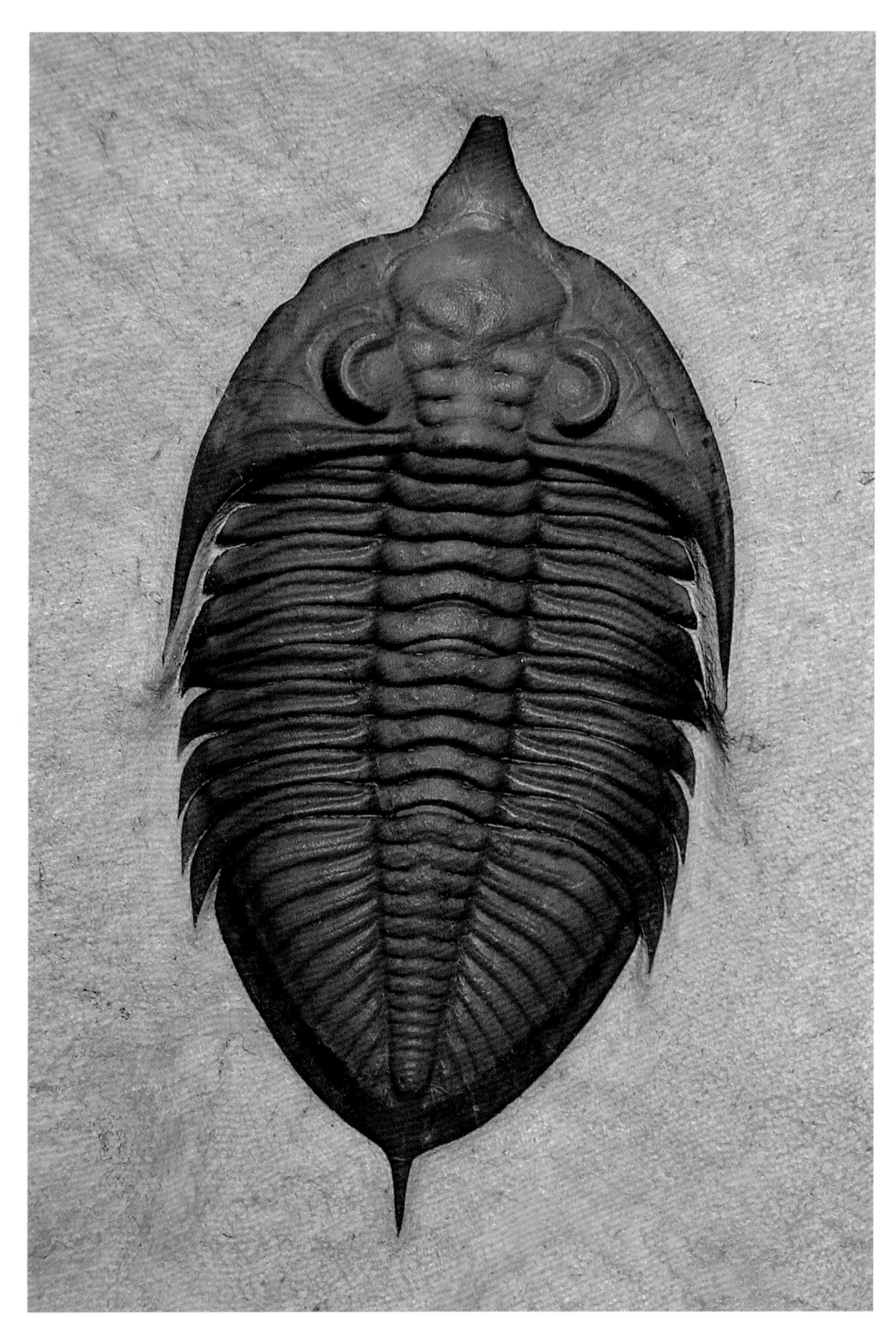

PLATE 149. *Huntonia (Huntonia) lingulifer* (Ulrich & Delo) (96 mm). L. Devonian, Haragan Formation, Coal County, Black Cat Mountain, Clarita, Oklahoma. Specimen courtesy of Bob Carroll, photographed by the author.

2.4 EASTERN NEWFOUNDLAND

The saga of my involvement with the trilobites found in the Middle Cambrian layers exposed in the Manuels River gorge of Conception Bay South, on the Avalon Peninsula of eastern Newfoundland, is narrated in the first and second editions of my trilobite books. Alas, something important to me, the color, has thus far been missing in the published pictures of these captivating fossils, with the exception of the cover of my second edition. Since 1993, when that edition appeared, I have been wishing to remedy this shortcoming, which limited the full visual impact of my findings for the majority of the Manuels River specimens. In the intervening 20 years, aided by the digital revolution, I am finally able to convey the reason for my excitement about the occasional color of trilobites.

It has been 40 years since my first encounter with the giant *Paradoxides davidis* of the Manuels River Formation. As mentioned in my first edition, my early preoccupation was taking steps toward the protection of this valuable fossil site, which was threatened by the dumping of rusty remains of cars and appliances and of boulders from the overlying construction sites. Although this obliteration was halted by local authorities, the formal protection of the entire Manuels River gorge as a nature preserve was enacted in 1992 with the constitution of the Manuels River Natural Heritage Society. This organization took charge of creating a complex of stairs, bridges, and walks along the Manuels River gorge to make it an attractive and accessible park that went well beyond the protection of the fossil exposures, enabling the visitors to enjoy the forested banks of the river, rich in exclusive wilderness flora and inspiring vistas. It has only been in the last two years that I was allowed to resume some fossil excavation, this time with the invaluable help of several board members of the society and of a dedicated paleontology student, Greg Froude. The society has recently undertaken the task of constructing a permanent visitor and educational center, with financial help of the local government, Hibernia, the Rotary Club of Avalon Northeast, and private donors. This new facility, named the Manuels River Hibernia Interpretation Centre, now houses much of my published and still unpublished personal trilobite collection, which, in addition to the Manuels River material, also contains many specimens of the same geological formations found in the quarries at the Red Bridge Road of Kelligrews.

The geological and paleontological significance of the trilobite fauna extracted from the strata exposed in the banks of the Manuels River are discussed in a professional paper that I coauthored with my longtime friend and mentor Jan Bergström (Bergström ad Levi-Setti 1978). Our novel findings, updating an early description in my 1975 trilobite book, are also illustrated in the 1993 second edition. Short of repeating the whole story, I prefer to highlight just what makes the Manuels River trilobites such a valuable and unique fossil record.

Striking is the undistorted preservation of the trilobite remains, either the exuviae shed in the molting process or the occasional occurrence of complete articulated individuals, presumably subjected to rapid burial while still alive. The only deformation is a moderate vertical compression under the weight of the overlaying shale strata, which, however, does not compromise the fine details of the surface ornamentation of the carapaces, which is best preserved in their external impressions. The diagenetic yellow coating of limonite, mentioned in the introduction, occurs frequently in these trilobites, providing unparalleled color contrast between fossil and embedding shale, as shown in this section.

This undisturbed taphonomy is exceptional considering the tectonic upheaval that brought these strata to the Avalon Peninsula. Indeed, these Middle Cambrian strata were originally deposited while connected with the European tectonic plate, as proven by the presence at Manuels of the same trilobite fauna, among which is the dominant giant *Paradoxides davidis,* otherwise exclusively found on the western coast of Wales on the shore at St. Davids in Pembrokeshire (hence the species name *davidis* instituted by Salter in 1863). Thus, the Manuels trilobites provide irrefutable evidence of continental drift, which brought the Avalon together with maritime North America away from the vicinity of Europe to the American side of the Atlantic after the breakup of the supercontinent named Pangaea. There is, however, a difference between the late taphonomies of the Manuels and the Wales fauna: while the Manuels depositional strata somehow floated undisturbed during and after the crunch of Pangaea, the Wales strata suffered dramatic tilting and tectonic shear that deformed the preserved trilobites into stretched or skewed shapes. Examples of the latter are shown in Bergström and Levi-Setti (1978) and in plate A3 of my second edition.

In addition to the above distinctive characters, the *Paradoxides* fauna of the Manuels River has provided us with a rare snapshot of evolution in action. This transpired in the detailed study reported in Bergström and Levi-Setti (1978), touching upon the genetics of population biology and evolution, all contained in three meters of strata explored layer by layer. What we found reminds me of Dukas's *The Sorcerer's Apprentice* as illustrated in Walt Disney's *Fantasia*. The Sorcerer is, in our case, the genetic mechanism that maintains the characteristics of an eukaryotic species constant over many generations in a large assemblage of individuals that we shall call the ancestral gene pool. If a small portion of this gene pool may become separated from the Sorcerer in a so-called peripheral isolate, the Sorcerer's Apprentice here, the latter may attempt to manifest its independence with gene mutations that may locally establish themselves, yielding individuals where some morphological character may differ from those of the Sorcerer (what is called phenotypic substitution). The reverse may also occur: the Sorcerer returns to take over the Apprentice in its act of rebellion and restores the original order—in our case, this happens if the peripheral isolate may reestablish contact with the ancestral gene pool. Such alternation may occur via sequential changes of a shallow sea level (regression and transgression in geological terms). We actually discovered three instances of such occurrence within three meters of the *Paradoxides* beds examined at Manuels. We named three mutants (the Apprentices), exclusively found in distinct layer bands, as subspecies of *Paradoxides davidis* (the Sorcerer or ancestral gene pool, now called *Paradoxides davidis davidis*). One of the mutants, with pleural spines shorter than the ancestor, was called *Paradoxides davidis brevispinus;* another, with a pygidium distally broader than the tapering one of the ancestor, was called *Paradoxides davidis trapezopyge;* and one with rectangular pygidium, found only at the interface between the other two, was named *Paradoxides davidis intermedius*. Eventually, a veritable daughter species appeared, but it was found only in Scandinavia (the Baltic province was nearby at that time) where all the above genetic mutant characters were preserved, named *Paradoxides forchammeri,* as hypothesized in Bergström and Levi-Setti (1978).

Unexpectedly, my treasure hunt in search of the colorful *Paradoxides* at Manuels has yielded information not only about continental drift but also about eukaryotic evolution, that is, the struggle that a daughter species may be subjected to in order to become a stable progeny and the fact that speciation does occur in a stepwise manner, much as postulated in the *punctuated equilibria* model of Gould and Eldredge (1977). Indeed, there is much more than meets the eye in the Manuels River trilobites.

PLATE 150. *Paradoxides davidis davidis* Salter (Bergström & Levi-Setti) (277 mm). M. Cambrian, Manuels River Formation, west side of Manuels River, Manuels, Newfoundland. RLS coll. This specimen was previously seen in a black-and-white image of the second edition of my book (Levi-Setti 1993, plate A2). This giant specimen is a rare representative of the ancestral phenotype mentioned in the introduction to the present section of this book, identical to that first discovered by Salter in 1863 on the West coast of Wales at St. David's in Pembrokeshire. The complete carapace suggests burial of a live individual, not a molt. When first extracted from the shale beds, it was covered with a bright layer of limonite, subject to being easily brushed off. A distinguishing characteristic of this subspecies is the distally tapered pygidium.

PLATE 152. *Paradoxides davidis davidis* Salter (Bergström & Levi-Setti) (117 mm). M. Cambrian, Manuels River Formation, west side of Manuels River, Manuels, Newfoundland. Matteo Levisetti coll. This complete specimen was extracted from the same shale level that yielded the slab shown in the preceding plate.

PLATE 151. *Paradoxides davidis davidis* Salter (Bergström & Levi-Setti) (vertical image length 220 mm). M. Cambrian, Manuels River Formation, west side of Manuels River, Manuels, Newfoundland. RLS coll. Now at MRHIC, Manuels. In the smaller upper molt, the cranidium has lifted away, however, leaving in place the other cephalic tergites and exposing the hypostoma attached to the rostral plate. In the underlying larger individual, the posterior third of the right thoracic pleurae are missing, not as a result of postburial damage.

PLATE 154. *Paradoxides davidis davidis* Salter (Bergström & Levi-Setti) (horizontal image length 21.0 cm). M. Cambrian, Manuels River Formation, west side of Manuels River, Manuels, Newfoundland. RLS coll. Now at MRHIC, Manuels. Coquina of molted cranidia, probably accumulated by shoreside wave motion, covering vast shale layers.

PLATE 153. *Paradoxides davidis davidis* Salter (Bergström & Levi-Setti) (66 mm). M. Cambrian, Manuels River Formation, west side of Manuels River, Manuels, Newfoundland. RLS coll. Now at MRHIC, Manuels. Exuviae of a juvenile individual that was previously illustrated in a black-and-white photograph (Levi-Setti 1993, plate A6). The last two thoracic segments, together with the attached pygidium, are displaced forward, overlapping the 13th to 15th segment. The cephalon is slightly tilted leftwise, and the crushed glabella exposes the underlying hypostoma.

PLATE 155. *Paradoxides davidis trapezopyge* Bergström & Levi-Setti (183 mm). M. Cambrian, Manuels River Formation, east side of Manuels River, Manuels, Newfoundland. RLS coll. Now at MRHIC, Manuels. External impression of complete exuviae, as suggested by the tilt of the cephalon with attached two anterior thoracic segments. The steinkern of this specimen is now at the Natural History Museum in Geneva, Switzerland. This trilobite is one of the mutant subspecies of *P. davidis* described in the introduction to this section (see Bergström and Levi-Setti 1978). It is characterized by the spatulated, trapezoidal shape of its pygidium, which differs from the distally tapered shape of the ancestor species illustrated in the preceding plates. It appears exclusively in a 37-cm section of the sequence of exposed layers.

PLATE 156. *Paradoxides davidis trapezopyge* Bergström & Levi-Setti (185 mm). M. Cambrian, Manuels River Formation, east side of Manuels River, Manuels, Newfoundland. RLS coll. This specimen was previously shown in a black-and-white photograph (Levi-Setti 1993, plate A4). The external impression of a hypostoma, possibly belonging to the same individual, appears superposed to the posterior left side of the thorax.

PLATE 157. *Paradoxides davidis trapezopyge* Bergström & Levi-Setti (179 mm). M. Cambrian, Manuels River Formation, east side of Manuels River, Manuels, Newfoundland. RLS coll. This specimen was also previously shown in a black-and-white photograph (Levi-Setti 1993, plate A4). It represents exuviae in the early stages of the molting process. Both free cheeks are released and displaced from their original setting; the right one is overturned. The hypostoma, also displaced, appears under the crushed cranidium. This specimen was found in 1974 by my son Emile, age nine, while splitting a large block of shale just extracted from the wall of sediments.

PLATE 158. *Paradoxides davidis trapezopyge* Bergström & Levi-Setti (horizontal image width 461 mm). M. Cambrian, Manuels River Formation, east side of Manuels River, Manuels, Newfoundland. RLS coll. Now at MRHIC, Manuels. The two interlocking slabs are fragments of a very large shale bed located at the bottom of the *P. d. trapezopyge* layers, literally covered with complete large individuals.

PLATE 159. *Paradoxides davidis intermedius* Bergström & Levi-Setti (175 mm). M. Cambrian, Manuels River Formation, east side of Manuels River, Manuels, Newfoundland. RLS coll. Now at MRHIC, Manuels. External impression. This subspecies of *P. davidis* occurs exclusively in a thin shale layer that separates the overlaying beds of *P. davidis davidis* from the underlying beds of *P. davidis trapezopyge*. A distinguishing characteristic is the parallel-sided rectangular pygidium.

PLATE 160. *Eccaparadoxides eteminicus* Matthew (153 mm). M. Cambrian, Fossil Brook Member, Upper Chamberlain's Brook Formation at Red Bridge Road second quarry, Kellygrews, Conception Bay South, Newfoundland. Matteo Levisetti coll. Exuvia missing the free cheeks. Well-known correlated strata containing this type of trilobite occur in New Brunswick.

PLATE 162. *Anopolenus henrici* Salter (75 mm). M. Cambrian, Manuels River Formation, *Paradoxides davidis* beds on west bank of Manuels River, Manuels, Newfoundland. RLS coll. Gift by Jan M. Chabala. Now at MRHIC, Manuels. An image of this specimen was previously published in black and white (Levi-Setti 1993, plate 95). This trilobite was originally discovered in Wales, UK. One of the free cheeks, released in molting, with its long slender genal spine, overlaps the thorax.

PLATE 161. *Eccaparadoxides pusillus* (Barrande), synonym of *Paradoxides rugulosus* Corda (Snajdr 1958) (53 mm). M. Cambrian, Manuels River Formation, *Paradoxides davidis* beds at Red Bridge Road lowest quarry, Kellygrews, Conception Bay South, Newfoundland. RLS coll. Now at MRHIC, Manuels.

PLATE 163. *Conocoryphe terranovicus* Resser (46 mm). M. Cambrian, Fossil Brook Member, Upper Chamberlain's Brook Formation at Red Bridge Road second quarry, Kellygrews, Conception Bay South, Newfoundland. Matteo Levisetti coll. The cranidium surface is finely granulated, the pygidium is displaced.

PLATE 164. *Ctenocephalus howelli Resser* (sagittal length 12 mm). M. Cambrian, Manuels River Formation, *Paradoxides davidis* beds on east bank of Manuels River, Manuels, Newfoundland. RLS coll. Now at MRHIC, Manuels. This cranidium is part of a large shale slab containing separated tergites of most of the fauna represented at this level. To be noted is the remarkable tuberculated surface of this specimen.

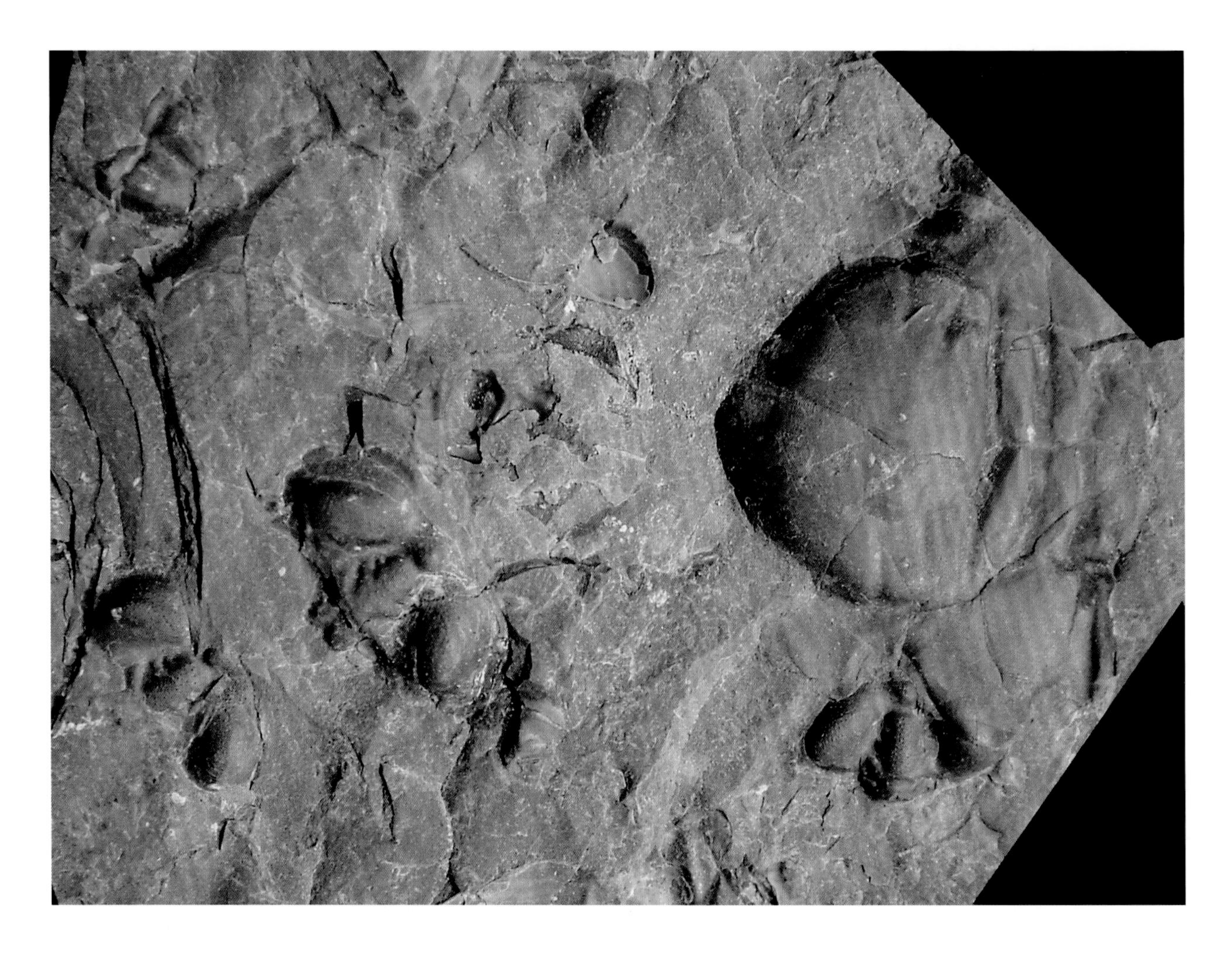

PLATE 165. *Elyx matthewi* Hutchinson (horizontal image width 11.0 cm). M. Cambrian, Fossil Brook Member, Upper Chamberlain's Brook Formation at Red Bridge Road second quarry, Kellygrews, Conception Bay South, Newfoundland. RLS coll. Now at MRHIC, Manuels. Several small cranidia of this trilobite; complete carapaces were not found. Typically seen in thin shale layers decorating large limestone blocks. On the right is a cranidium of *Eccaparadoxides eteminicus* Matthew.

PLATE 166. *Eodiscus punctatus* Salter (horizontal image width 30 mm). M. Cambrian, Manuels River Formation, *Paradoxides davidis* beds at Red Bridge Road lowest quarry, Kellygrews, Conception Bay South, Newfoundland. RLS coll. Coquina of disarticulated tergites; no complete specimen was found. Note minute surface ornamentation of these small trilobites.

PLATE 167. Another area of the sample shown in plate 166, containing punctuated cranidia of *Bailiella* sp. The densely populated thin shale slabs containing this fauna measured in excess of 30 cm.

PLATE 169. *Diplagnostus planicauda* (Tullberg) (horizontal image width 32 mm). M. Cambrian, Manuels River Formation, *Paradoxides davidis* beds on west bank of Manuels River, Manuels, Newfoundland. RLS coll. Interspersed with this predominant type are occasional exuviae of other agnostids.

PLATE 168. *Tomagnostus perrugatus* Groenwall (horizontal image width 62 mm). M. Cambrian, Manuels River Formation, *Paradoxides davidis* beds on west bank of Manuels River, Manuels, Newfoundland. RLS coll. The disarticulated exuviae of these small agnostid trilobites are found selectively collected over large shale surfaces.

PLATE 170. Montage of two complete agnostids. On the upper left is an example of *Ptychagnostus punctuosus* (Angelin) (9 mm); on the lower right is *Triplagnostus gibbus* (Linnarson) (8 mm). M. Cambrian, Manuels River Formation, *Paradoxides davidis* beds on west bank of Manuels River, Manuels, Newfoundland. RLS coll.

2.5 UNITED KINGDOM

My visits to the United Kingdom dealt with two stages of my scientific career. In the first stage, back in the early 1950s at the University of Milan, Italy, my focus was on the physics of elementary particles, when major discoveries were to be made. In 1952, I was invited to present some of my findings at the Royal Society in London, where the lecture room where Newton taught about gravity was rather intimidating. Then came a collaboration with colleagues at the University of Bristol, where Nobel-winning discoveries in my field of research had been made a few years earlier. With my Italian upbringing, I found some discomfort in adapting to British cuisine, the habit of driving on the "wrong" side of the road, and the coin-operated heating devices, which roasted your front, while prompting a cold draft on your backside.

The second stage, some 20 years later, dealt with a radical shift of my interests. I abandoned elementary particles and high-energy physics in favor of developing and using a novel analytical instrument, a high-resolution scanning ion microprobe, to investigate very small structures of materials and biology. At the same time, I rekindled an early interest in invertebrate paleontology, a sought-after recreational escape from physics. Little did I know that physics would pervade even this attempted escape on my part. My passion being trilobites, with their tantalizing compound eyes with calcite lenses in vivo, I soon found out that these primordial creatures made use of little known but fundamental functions of optics. This came about when I attended a trilobite conference in Oslo, Norway, in 1972, where I was exposed to a lecture by Euan N. K. Clarkson, who described peculiar doublet structures in the eye lenses of a particular class of trilobites. As described in detail in my earlier trilobite books, and which is touched upon again in this book in the section devoted to images of trilobite eyes, I found out that these structures corrected the spherical aberration of the eye lenses according to the optical canons that Huygens and Descartes set forth back in the 17th century, and which are still in use today in camera lenses as well as in the Hubble telescope. There went my escape from physics into paleontology. Thus, a collaborative seminal paper was published jointly (Clarkson and Levi-Setti 1975), followed by prize-winning publications over years to come, about the earliest visual systems known.

During my repeated visits to Edinburgh to work with Euan Clarkson, I was exposed to Scottish cuisine, the ultimate challenge to my usual nourishment. Nevertheless, aspects of Edinburgh were most appealing, among which were the many beautiful rose gardens flanking my daily walks to the Grant Institute of Geology. Together, we went digging for trilobites in the nearby Pentland Hills, and one rainy, cold day, we took a field trip to Girvan, on the western Scottish coast, after a terrifying drive in a London friend's Jaguar. There we were drenched searching for *Cyclopyge* trilobites on slippery rocks between waves and, more fruitfully, on an inland exposure that yielded some beautiful Ordovician dalmanitid *Calyptaulax* with large schizochroal eyes (see plate 24 of my second edition). However, my exposure to the chill made me familiar with the British National Health, to take care of a galloping bronchitis with appropriate linctus. At the time of my early visits to Edinburgh, I had been digging for the giant Middle Cambrian trilobites *Paradoxides davidis* Salter on the slopes of the Manuels River gorge in the Avalon Peninsula of eastern Newfoundland. This trilobite had originally been discovered at Porth y Rhaw, St. Davids, Wales (hence its name). Curious to compare my specimens from Newfoundland with those from Wales, Euan Clarkson and I chased in various museums examples from the Wales type locality. He also contacted on my behalf Richard A. Fortey at the British Museum of Natural History in London, who kindly supplied me with a mold of BM 1, the first type specimen in the British Museum Geology collection. The correspondence of the specimens from the two locales turned out to be perfect, in spite of the tectonic shear present in most examples from Wales. The continental drift implication of this transcontinental correspondence is discussed in detail in my earlier trilobite books and in the paper by Bergström and Levi-Setti (1978).

Wales became an attractive hunting ground for other trilobite species on a visit to meet Robert J. Kennedy of Birmingham, a most knowledgeable amateur trilobite collector. In a very limited dwelling space, he had amassed a most impressive collection of British trilobites. He kindly drove me on a memorable tour of Welsh trilobite sites, hidden among riverbeds crossing farm fields, where permission to trespass and dig for fossils was always graciously granted by intrigued landowners. I learned that one could spend comfortable nights in rooms that were often available above country pubs, which on one occasion even offered impressive Italian cuisine. Favorite destinations became localities near Llandrindod Wells and a quarry near Pencerrig Lake, Llanewedd, Powis, where practically every rock fragment and sometimes large slabs of shale yielded colorful specimens of the L. Ordovician *Ogygiocarella debuchii.* We also spent time in a quarry near the top of Gilvern Hill, not far from Builth-Wells, Powis, digging for the large M. Ordovician *Ogyginus corndensis.* Seen from the top of the hill, a sequence of similar hills surrounded our perch, as far as the horizon, with nobody in sight. Suddenly, a jogger in shorts appeared from nowhere trotting along the crest, asking us if this was the way to Llandrindod-Wells. We really could not be of much help, but he reappeared on top of the next hill, with a confident gait, well before the advent of GPS devices. We regarded this unexpected encounter rather amusing and pondered why Welsh names carry a multitude of seemingly redundant double consonants.

PLATE 171. *Ogyginus corndensis* (Murchison) (11.5 cm). L. Ordovician, Gilvern Hill, Builth-Wells, Powis, Wales, UK. RLS coll.

PLATE 172. *Ogyginus corndensis* (Murchison) (19 mm). L. Ordovician, Gilvern Hill, Builth-Wells, Powis, Wales, UK. Juvenile individual.

PLATE 173. *Bettonia chamberlainii* Elles (6 mm). L. Ordovician, Gilvern Hill, Builth-Wells, Powis, Wales, UK. RLS coll.

PLATE 174. *Ogygiocarella debuchii* (Brongniart) (63 mm). M. Ordovician, stream section near Pencerrig Lake, Llanelwedd, Powis, Wales, UK. RLS coll.

PLATE 176. *Ogygiocarella angustissima* (Brongniart) (16 mm). M. Ordovician, Landeilo Series, Meadowtown beds, Meadowtown, Shropshire, UK. RLS coll. Gift by Robert J. Kennedy.

PLATE 177. *Flexicalymene arcuata* (Price) (17 mm). U. Ordovician, Ashgill Series, Dolhir beds, Cynwid Forest, North Wales, UK. RLS coll. Gift by Robert J. Kennedy.

PLATE 175. *Cnemidopyge nuda* (Murchison) (34 mm). M. Ordovician, stream section, Pencerrig, Powis, Wales, UK. RLS coll. Gift by Robert J. Kennedy.

PLATE 178. *Calymene blumenbachi* Brongniart (89 mm). Silurian, Wenlock Limestone Formation, Wren's Nest, Dudley, West Midlands, UK. RLS coll. This is a brass cast of the famous "Dudley Bug" from Nature's Own, Nederland, Colorado.

2.6 RUSSIA

Since I have never been to Russia, I was fortunate to meet Arkadiy Evdokimov at the annual Tucson Gem and Mineral Show. This young Russian paleontologist was in charge of the magnificent exhibits of the Ordovician trilobites from the St. Petersburg area collected and prepared by the Petersburg Paleontological Laboratory. These trilobites are extracted from a narrow belt of limestone quarries and outcrops located south of St. Petersburg and extending close to the Baltic coastline from the Estonian border on the west to the Ladoga Lake on the east. Encased in whitish fine-grained limestone, the trilobites are beautifully preserved in their full tridimensionality with their carapaces replaced by tan-colored translucent calcite. The removal of the matrix from delicate free-standing spines and surface ornamentation by airbrasive techniques is a remarkable feat that distinguishes the painstaking preparation methodologies practiced by this laboratory.

In my quest for photogenic trilobite specimens to purchase and bring home to photograph at leisure for this book, I was confronted with the, for me inaccessible, cost of the most intriguing Russian trilobites in Arkadiy's exhibit. Thus I asked him for permission to take some pictures of his masterpieces, after placing them in an improvised studio set-up. I was pleasantly surprised by Arkadiy's offer to send me some of his photographs, taken in the most appropriate lighting set-ups. After the show was over, I received in the mail from him a CD with 300 high-resolution shots from his collections for me to choose from for my book. I could not pass up this splendid opportunity, and present here a section devoted mostly to his contribution of exquisite photos, where trilobite's architectures and postures are viewed with often lyrical composition.

PLATE 179. *Asaphus lepidurus* Nieszkowski (79 mm). L. Ordovician, St. Petersburg Region, Russia. RLS coll. The correct identification of the species of this genus is somewhat problematic in view of subtle differences among a multitude of compatible attributions.

For well over a century, the trilobites of the St. Petersburg deposits have attracted the attention of paleontologists worldwide, who described their findings in an enormous trove of publications. The work of the Petersburg Paleontological Laboratory has been publicized in numerous web entries, describing all aspects of their work, from the quarrying of the host exposures to the steps involved in their preparation as well as the final exhibit-quality outcome. This effort was accompanied by professional descriptions of the trilobite's morphology, their historical identification, and their classification. The work of many years has now been collected in a beautiful book entitled *Ordovician Trilobites of the St. Petersburg Region, Russia* by V. Klikushin, A. Evdokimov, and A. Pilipyuk, published in 2009 by Petersburg Paleontological Laboratory/Griffon Enterprises Inc. Given the rich assortment of photographs placed at my disposal by Arkadiy Evdokimov, I have attempted, whenever possible, to select trilobite views different from those contained in that book.

Turning now to the circumstances that prevailed during the Ordovician period when the rich trilobite fauna of the St. Petersburg area left such an exceptional record of its life, it is believed that what is now eastern Russia was a shallow inland sea some 20 to 100 meters deep. Evidence suggests that this basin was repeatedly in contact with the open sea on the western side, while becoming a "peripheral isolate" (in the language of population biology) when such contact was interrupted. In this environment, genetic diversification of morphologies is fostered, as suggested by the presence in the fossil record of 91 genera, comprising 209 species, assigned to 31 families, identified and described in *Ordovician Trilobites of the St. Petersburg Region, Russia.* The evolutionary history of these trilobites, from progenitors to descendants, is written in the sequence of limestone layers and is most prolific on the so-called Aseri horizon, a 20-meter-thick band along the Wolchov River. In the peripheral isolate environment, mutations and adaptive strategies have a better chance to survive, inclusive of the development of exotic and extravagant morphologies, such as those encountered in some Asaphidae, Cheiruridae, Lichidae, and Odontopleuridae. Several of these are represented in my selection from Arkadiy's CD, including a few from other sources, but all prepared by the Petersburg Paleontological Laboratory.

PLATE 180. *Illaenus tauricornis* Kutorga (60 mm, left side of image) and *Asaphus kowalevskii* Lowrow (65 mm, larger individual at right side of image). M. Larnvinian (L.-M. Ordovician), St. Petersburg Region, Russia. Specimen courtesy of Dave Douglass, photographed by the author. In *Asaphus kowalevskii,* the eyes are located at the tip of long stems. This feature suggests periscopic vision when the body of the trilobite was buried in sand or mud, a survival precaution.

PLATE 181. *Subasaphus platyurus* (Angelin) (135 mm). M. Ordovician, St. Petersburg Region, Russia. Specimen courtesy of Nord Fossils, photographed by the author.

PLATE 182. *Megalaspidella isvosica* Balashova (82 mm). L. Ordovician, St. Petersburg Region, Russia. Photograph courtesy of Arkadiy Evdokimov.

PLATE 183. *Megalaspidella* sp. (110 mm). L. Ordovician, St. Petersburg Region, Russia. Photograph courtesy of Arkadiy Evdokimov. The correct species identification of this specimen is uncertain, since the overall morphology of the carapace corresponds to that of *Megalaspidella isvosica* Balashova, with the exception of the anterior tip of the cranidium, which is rounded and similar to that of *Megalaspidella obtusa* Schmidt.

PLATE 184. *Pliomera fisherii* (Eichwald) (43 mm). L. Ordovician, St. Petersburg Region, Russia. RLS coll.

PLATE 185. *Paraceraurus exsul* (Beyrich) (79 mm). M. Ordovician, St. Petersburg Region, Russia. RLS coll. This is one of the most aesthetically pleasing trilobites, delicately prepared at the Petersburg Paleontological Laboratory.

PLATE 186. *Pandaspinapyga tumida* (Angelin) (56 mm). L. Ordovician, St. Petersburg Region, Russia. Photograph courtesy of Arkadiy Evdokimov.

PLATE 187. *Cybele panderi* Schmidt (52 mm). U. Larnvinian (L.-M. Ordovician), St. Petersburg Region, Russia. Specimen prepared at the Petersburg Paleontological Laboratory, courtesy of Dave Douglass, photographed by the author.

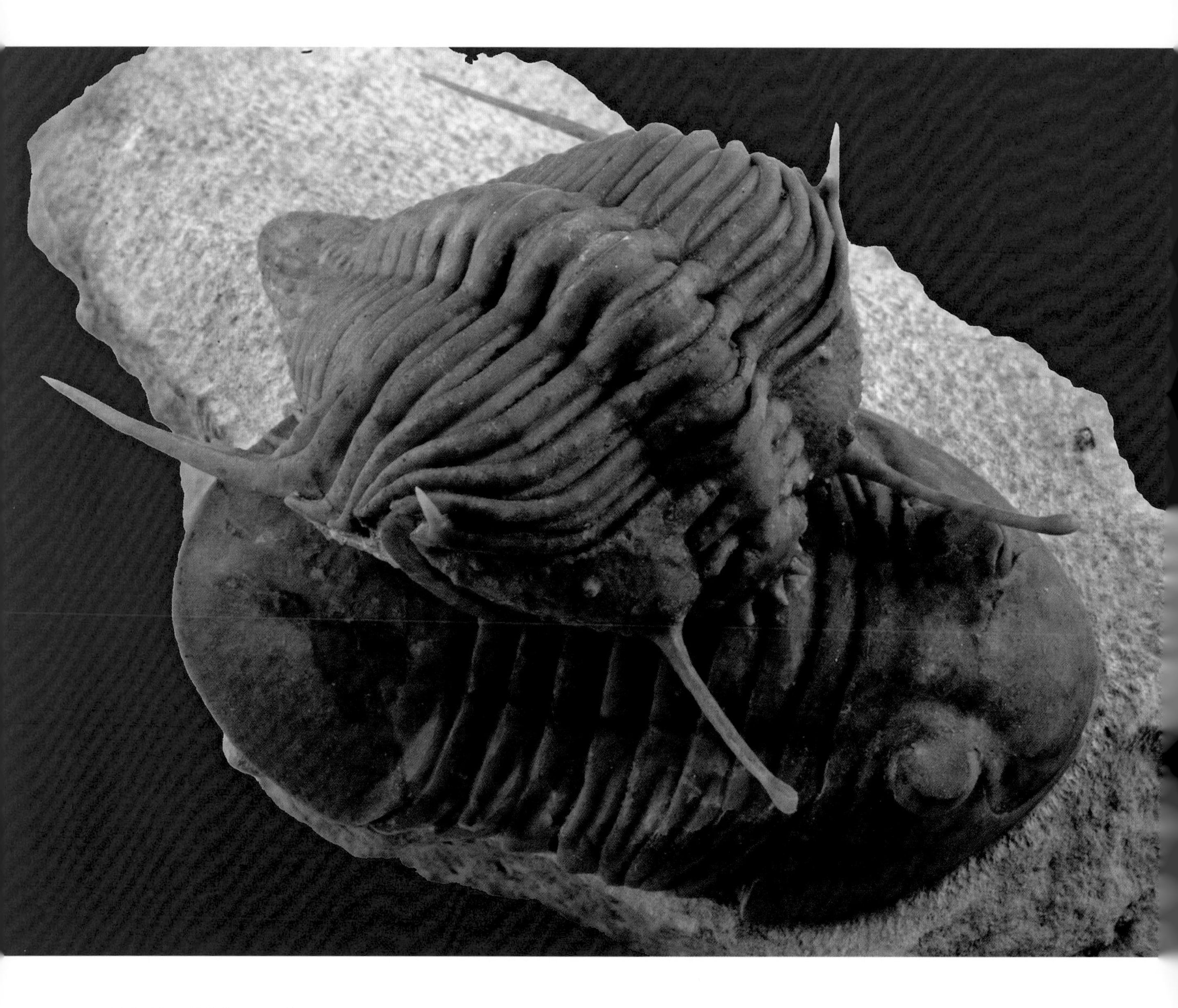

PLATE 188. *Cybele panderi* Schmidt overlapping the body of a specimen of *Asaphus plautini* Schmidt (98 mm). U. Larnvinian (L.-M. Ordovician), St. Petersburg Region, Russia. Photograph courtesy of Arkadiy Evdokimov.

PLATE 189. *Prochasmops praecurrens* (Schmidt) (54 mm). U. Larnvinian (L.-M. Ordovician), St. Petersburg Region, Russia. Photograph courtesy of Arkadiy Evdokimov.

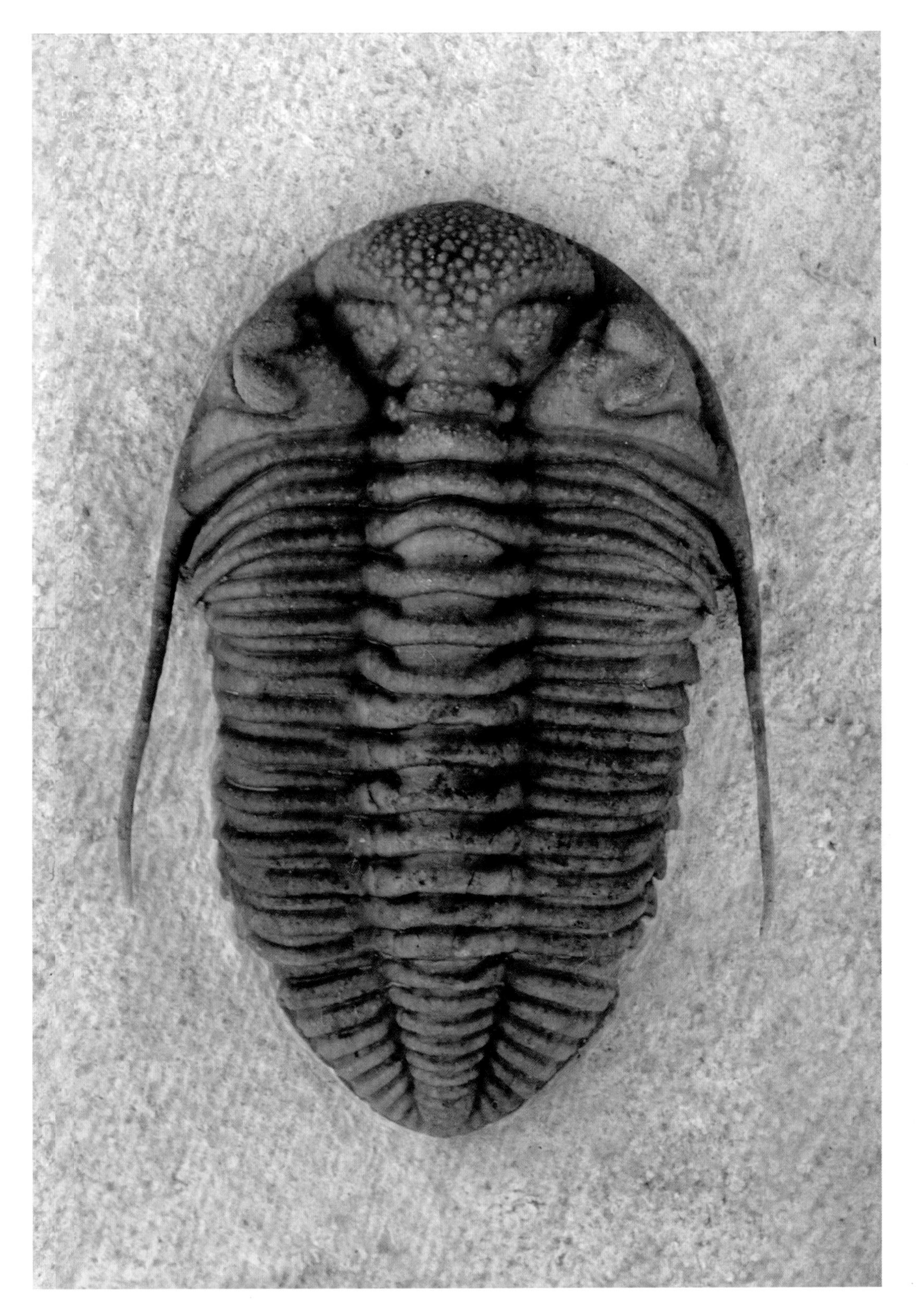

PLATE 190. Schizochroal eye of the preceding specimen. Photograph courtesy of Arkadiy Evdokimov.

PLATE 191. *Metopolichas huebnerii* (Eichwald) (85 mm). L. Landeilian (L.-M. Ordovician), St. Petersburg Region, Russia. Photograph courtesy of Arkadiy Evdokimov.

PLATE 193. *Hoplolichas tricuspidatus* (Beyrich). U. Larnvinian (L.-M. Ordovician), St. Petersburg Region, Russia. Photograph courtesy of Arkadiy Evdokimov. Lateral view of the specimen shown in the preceding plate. Note the schizochroal eye with few lenses.

PLATE 192. *Hoplolichas tricuspidatus* (Beyrich) (74 mm). U. Larnvinian (L.-M. Ordovician), St. Petersburg Region, Russia. Photograph courtesy of Arkadiy Evdokimov. Dorsal view of an extended specimen. The spectacular spinosity of the entire carapace and its exquisite preparation will be more evident in the following lateral view.

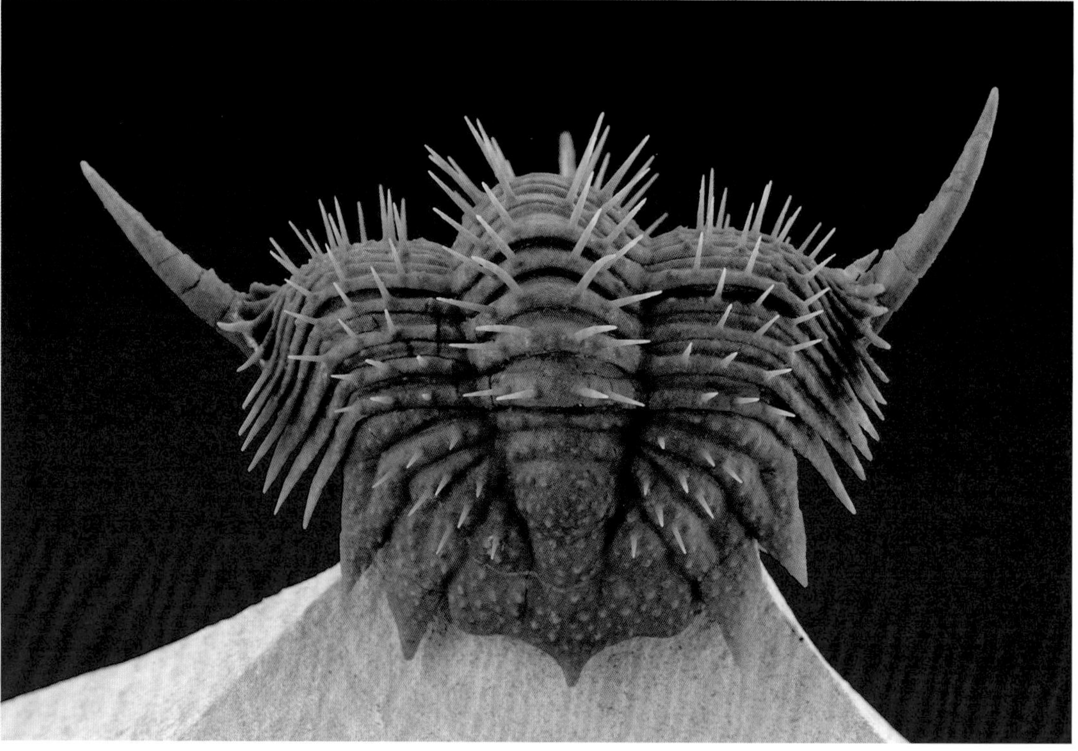

PLATE 196. *Hoplolichas plautini* Schmidt (57 mm). U. Larnvinian (L.-M. Ordovician), St. Petersburg Region, Russia. Specimen, prepared at the Petersburg Paleontological Laboratory, courtesy of Dave Douglass, photographed by the author. The pygidium and cranidium are covered with tubercles, larger in the latter.

PLATE 194. *Hoplolichas tricuspidatus* (Beyrich). U. Larnvinian (L.-M. Ordovician), St. Petersburg Region, Russia. Photograph courtesy of Arkadiy Evdokimov. Oblique frontal view of an enrolled specimen of the same species and similar size as that shown in the preceding two plates.

PLATE 195. *Hoplolichas tricuspidatus* (Beyrich). U. Larnvinian (L.-M. Ordovician), St. Petersburg Region, Russia. Photograph courtesy of Arkadiy Evdokimov. Posterior view of the specimen shown in the preceding plate.

PLATE 197. *Hoplolichoides conicotuberculatus* Nieszkowski (53 mm). M. Ordovician, St. Petersburg Region, Russia. Photograph courtesy of Arkadiy Evdokimov.

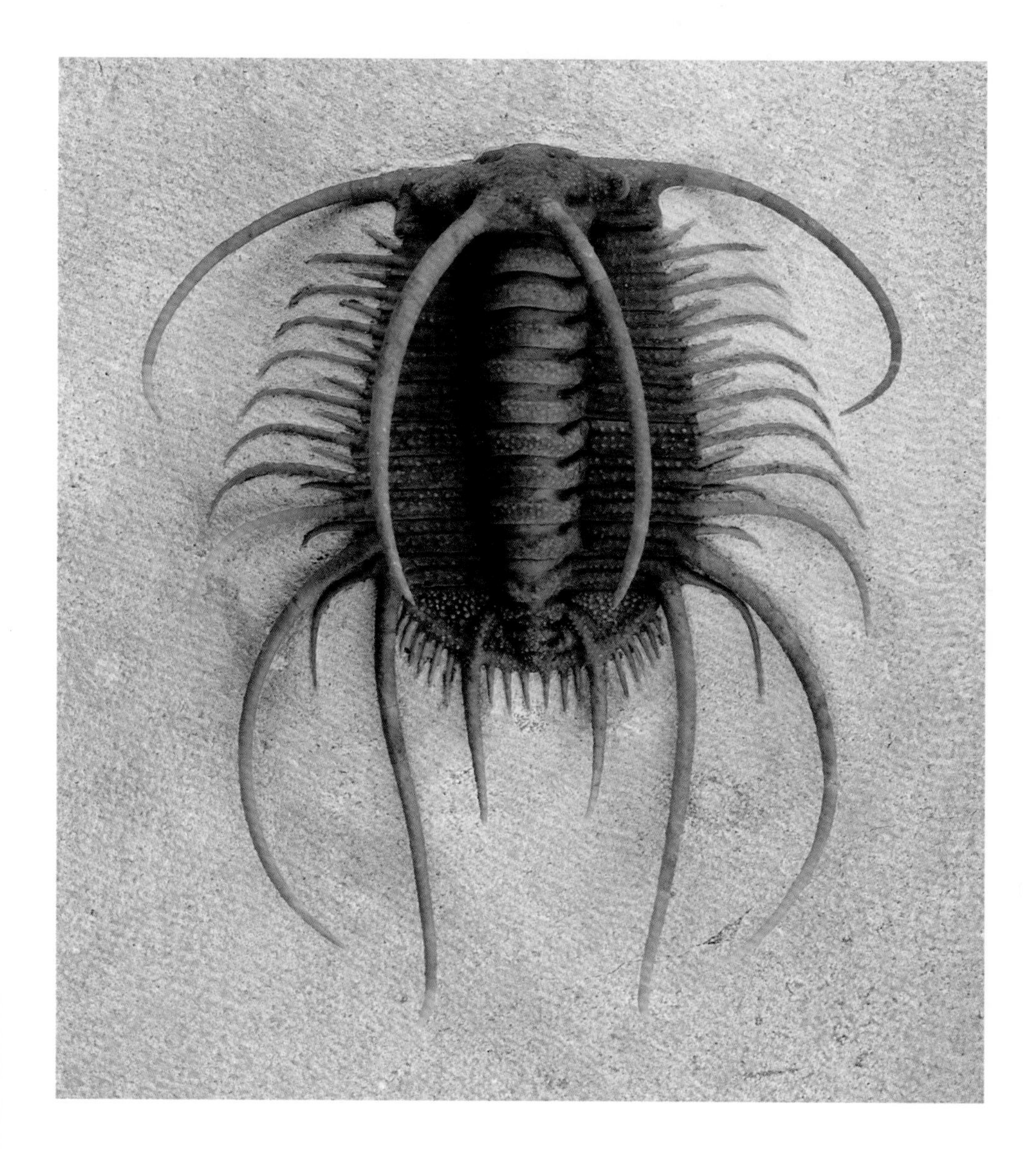

PLATE 198. *Boedaspis ensifer* Whittington & Bohlin (61mm). M. Ordovician, St. Petersburg Region, Russia. Photograph courtesy of Arkadiy Evdokimov. Dorsal view of extended specimen. The outstanding presentation of this Odontopleurid trilobite raises wonder and questions. The preparation of the freely standing delicate spines is nothing short of miraculous. What could have been the environmental pressures that motivated their evolution and function?

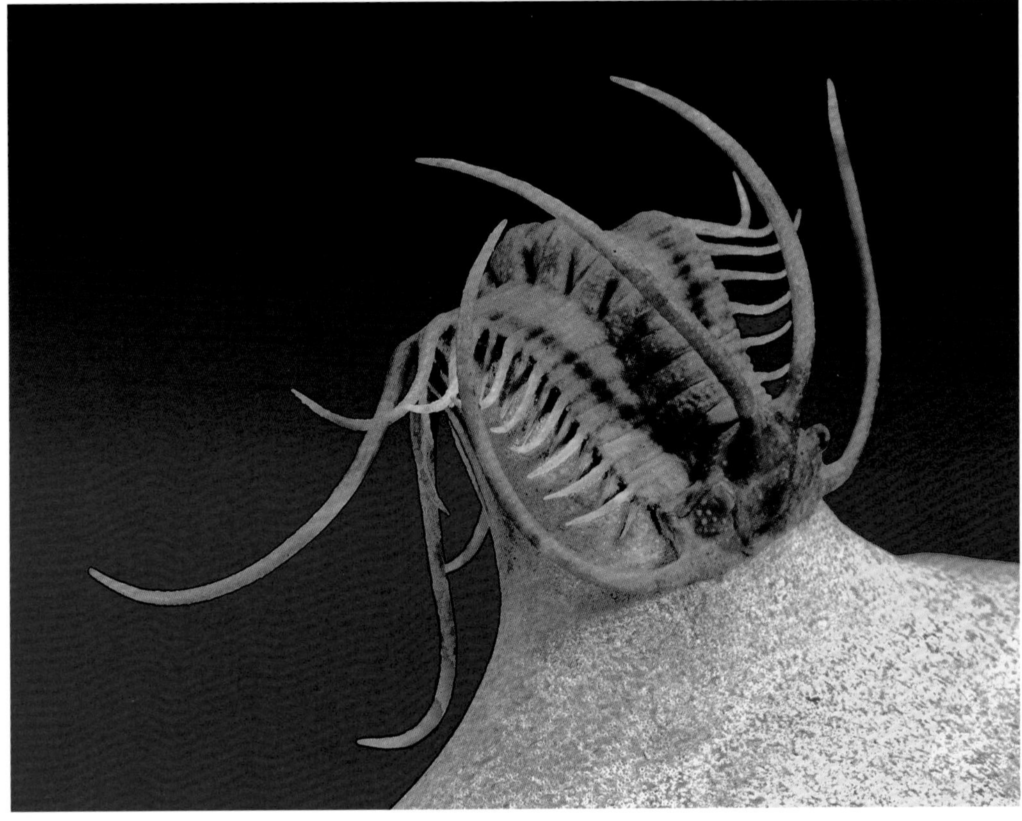

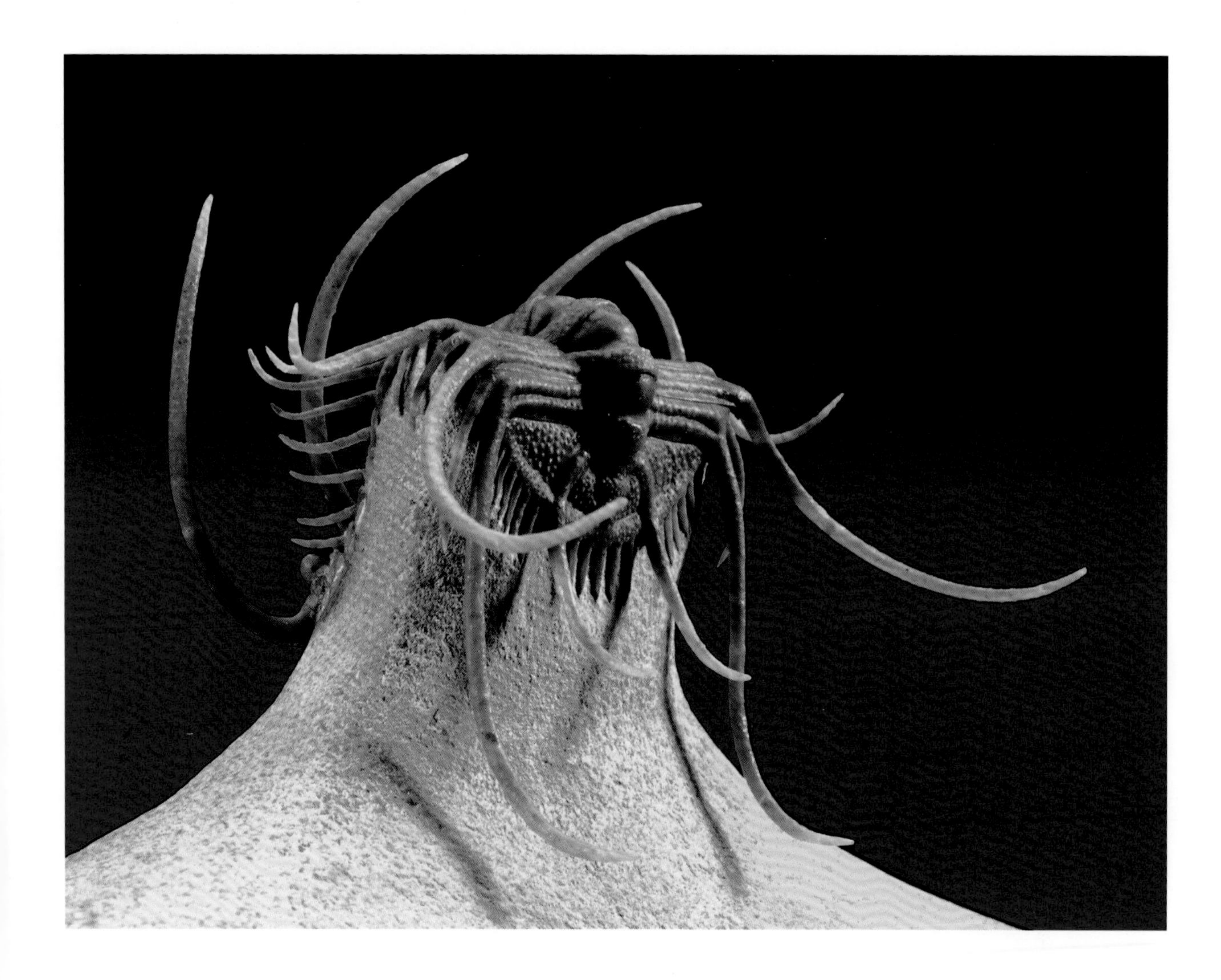

PLATE 201. *Boedaspis ensifer* Whittington & Bohlin. M. Ordovician, St. Petersburg Region, Russia. Photograph courtesy of Arkadiy Evdokimov. Posterior oblique view of the partially enrolled specimen shown in the preceding plate.

PLATE 199. *Boedaspis ensifer* Whittington & Bohlin. M. Ordovician, St. Petersburg Region, Russia. Photograph courtesy of Arkadiy Evdokimov. Frontal oblique view of the specimen shown in the preceding plate.

PLATE 200. *Boedaspis ensifer* Whittington & Bohlin. M. Ordovician, St. Petersburg Region, Russia. Photograph courtesy of Arkadiy Evdokimov. Frontal oblique view of a partially enrolled specimen. Here the elegant design of the ensemble of the freely standing spines acquires a lyrical quality.

2.7 TUCSON GEM AND MINERAL SHOW

Trilobite enthusiasts do not need Facebook to get in touch with each other and exhibit their precious, expertly prepared findings. Professional dealers from all over the world and their clientele converge on Tucson yearly during the first two weeks in February. During the first week, most exhibits take place in hotel rooms, their contents spilling onto front yards and sidewalks lining huge hotel plazas. Specimens are generally sold "keystone," namely, at 50% of regular price, in a friendly and attractive atmosphere. Other exhibits are set up in temporary tents and more sumptuous buildings. In the second week, the dealers transfer their wares to the main convention center in the middle of town.

Over many years, this event has enticed my curiosity to explore firsthand new hunting grounds in remote localities, as well as to appease my search for new, affordable specimens. More often, my camera remedied my limitations, helped by the courtesy of many exhibitors who allowed me to bring home the images of otherwise inaccessible material for my books. In the process, many new friendships enriched my quest. Rivaling museum exhibits, the spectacular size and perfection of the trilobite slabs extracted from the Moroccan sub-Sahara, covered by multitudes of complete individuals caught in their instant burial, always astonishes me. Often such slabs portray snapshots of the seafloor and its inhabitants, if they were provided with a mineralized exoskeleton that survived taphonomic obliteration, like the Ordovician brittlestars and the ubiquitous crinoids. Occasionally, the trilobite assemblages of complete individuals are so dense that one wonders whether they may represent transient mating episodes, not accumulations of randomly shed exuviae. In general, meter-sized slabs are collected in fragments, carefully reassembled in dusty ateliers, as I witnessed many times in Morocco. The removal of the top matrix encasing the trilobites was, in the early days, carried out by workers using nails as chisels, driven by handheld hammers. In recent years, pneumatic driven engraving points have upgraded the process drastically in speed and precision. Experienced preparers are now using airbrasive techniques such as miniature sandblasters to expose in their 3-D entirety delicate protruding spines and surface ornamentation of selected isolated specimens.

Unfortunately, several factors have conspired in fostering a prosperous industry of fossil reproductions, some innocuous if declared as such, others damaging the reputation of the trade if claimed as originals. This industry emerged in Morocco from the availability of rich fossil deposits in a region overflowing with destitute people desperate to find ways to make a living. Uneducated desert inhabitants and their children quickly learned to collect or excavate the valuable trilobite treasures to offer them for sale to avid foreigners. At first, they learned to repair damaged or missing details of the recovered carapaces with painted replacements. Once the basic anatomy of different genera was learned, they became adept at assembling parts from available incomplete specimens to create seemingly complete individuals, at times modified with added fantasized features. Finally, they perfected the ability to make faithful casts of genuine rare and valuable specimens, using materials simulating the original mineral composition and appearance of both fossil and rock matrix. These fake trilobites have become difficult to identify as such, even to the expert eye. Nowadays, the web offers innumerable readings about how to detect forgeries, after a seminal tutorial written for the Proceedings of the 2003 Munich Gems and Fossils Show (Burkhard and Bode 2003).

While roaming through the exhibits of Moroccan dealers in Tucson, the above warnings kept me alerted to spot trilobites that seemed to be identical repeats, the most compelling sign of forgeries. I did find a few remarkable examples, which I decided to show in this section, without revealing their source. However, the specimens that I chose to present in section 2.2 devoted to Morocco either were my findings or were obtained unprepared and then were prepared either by myself or by some of my more adept friends.

Aside from the overwhelming abundance of Moroccan trilobites, and the fortunate encounter with the Russian collections dealt with in section 2.6, the Tucson Show gave me the opportunity to have access to isolated, unusual trilobite specimens from other world provenances beyond my reach. This category includes specimens from China, Australia, and western Canada, not numerous enough to warrant separate sections in this book.

An unexpected honor was bestowed upon me in 2009 with the award of the prestigious Charles Sternberg Medal of the American Association of Applied Paleontology, for my contributions to paleontology. This remains a treasured memory of my visits to the Tucson Show.

REPUTABLE ORIGINALS OF MOROCCAN TRILOBITE ASSEMBLIES AND SINGLE SPECIMENS

PLATE 202. *Acadoparadoxides briareus* Geyer (average length of each trilobite ~300 mm). Lower to Middle Cambrian boundary, from a huge quarry at Tassemamt, Jbel Ougnate, Morocco. Slab exhibited at the Sahara Sea Collection, courtesy of Bill Barker, photographed by the author.

PLATE 203. *Dikelokephalina brechleyi* Fortey (average length of each trilobite ~270 mm). L. Ordovician, L. Fezouata Formation, Ouled-Slimane, Ain Chika, near Agdz, Morocco. Slab exhibited at the Sahara Sea Collection, courtesy of Bill Barker, photographed by the author. In a pertinent discussion by Corbacho and Vela (2010) and related references, finding assemblages of perfectly preserved complete individuals such as those illustrated here may imply sudden anoxia and consequent death due to lack of oxygen, creating a nonoxidizing environment, followed by immediate burial, perhaps caused by underwater volcanic events.

PLATE 204. *Megistaspis (Ekeraspis) hammondi* Corbacho & Vela (single, 245 mm) surrounded by many *Symphysurus* sp. aff. *Symphysurus palpebrosus* (Dalman) (each 42 to 50 mm). L. Ordovician, L. Fezouata Formation, central Anti-Atlas, Morocco. Slab exhibited at the Sahara Sea Collection, courtesy of Bill Barker, photographed by the author.

PLATE 205. *Metapilekia bilirata* Harrington (~65 mm each). L. Ordovician, L. Fezouata Formation, central Anti-Atlas, Morocco. Slab exhibited at the Sahara Sea Collection, courtesy of Bill Barker, photographed by the author. This slab also contains a specimen of *Megistaspis* sp.

PLATE 206. *Selenopeltis buchii* (Barrande) (average sagittal length ~150 mm). U. Ordovician, Tiouririne Formation (Ktaoua Group) near Erfoud, Morocco. Slab exhibited at the Sahara Sea Collection, courtesy of Bill Barker, photographed by the author. Several smaller Ordovician species are also present in this slab. See also similar assemblages of *Selenopeltis* in plate 46 of the section of this book devoted to Morocco.

PLATE 207. Another single *Selenopeltis buchii* (Barrande) amidst an assemblage of brittlestars, illustrated in plate 47 of this book. Slab exhibited at the Sahara Sea Collection, courtesy of Bill Barker, photographed by the author.

PLATE 208. A decorative repeat of a slab similar to that of plate 47 of this book, describing a view of the Ordovician seafloor populated by *Selenopeltis buchii* (Barrande) with starfish and brittlestars. Slab exhibited at the Sahara Sea Collection, courtesy of Bill Barker, photographed by the author.

PLATE 209. *Onnia superba* Bancroft (average length 14 mm). U. Ordovician, Ktaoua Formation, Bordj, Morocco. Portion of a slab exhibited at Sahara Overland, courtesy of Adam A. Aaronson, photographed by the author. Assemblages such as this may extend over several meters of sediment.

PLATE 211. *Platypeltoides magrebiensis* Rábano (average sagittal length ~200 mm). L. Ordovician, L. Fezouata Formation, Central Draa Valley, Morocco. Portion of giant slab exhibited at the Sahara Sea Collection in 2004, courtesy of Bill Barker, photographed by the author.

PLATE 210. *Asaphellus* sp. (average sagittal length ~160 mm). L. Ordovician, L. Fezouata Formation, Ouled-Slimane, near Agdz, Morocco. Portion of slab, seen from an angled view, exhibited in 2002, at the Sahara Sea Collection, courtesy of Bill Barker, photographed by the author. A frontal view of the very same slab appears on the website of the Royal Ontario Museum, Toronto, Ontario, Canada, where it now resides.

PLATE 212. Display window at the Sahara Sea Collection, exhibited in 2009, courtesy of Bill Barker, photographed by the author. All these specimens have been exquisitely prepared by expert preparators, mostly in Erfoud, Morocco.

PLATE 213. Detail of another display window at the Sahara Sea Collection, courtesy of Bill Barker, photographed by the author.

PLATE 214. Display window of Moroccan trilobites at Horst Burkhard, Minerals, exhibited in 2009, by courtesy of the dealer, photographed by the author. Some price tags can be deciphered, by amounts not surprising in view of the rarity of the specimen and the painstaking superb preparation of these spiny trilobites.

PLATE 215. Another display window at Horst Burkhard, Minerals, exhibited in 2009, by courtesy of the dealer, photographed by the author.

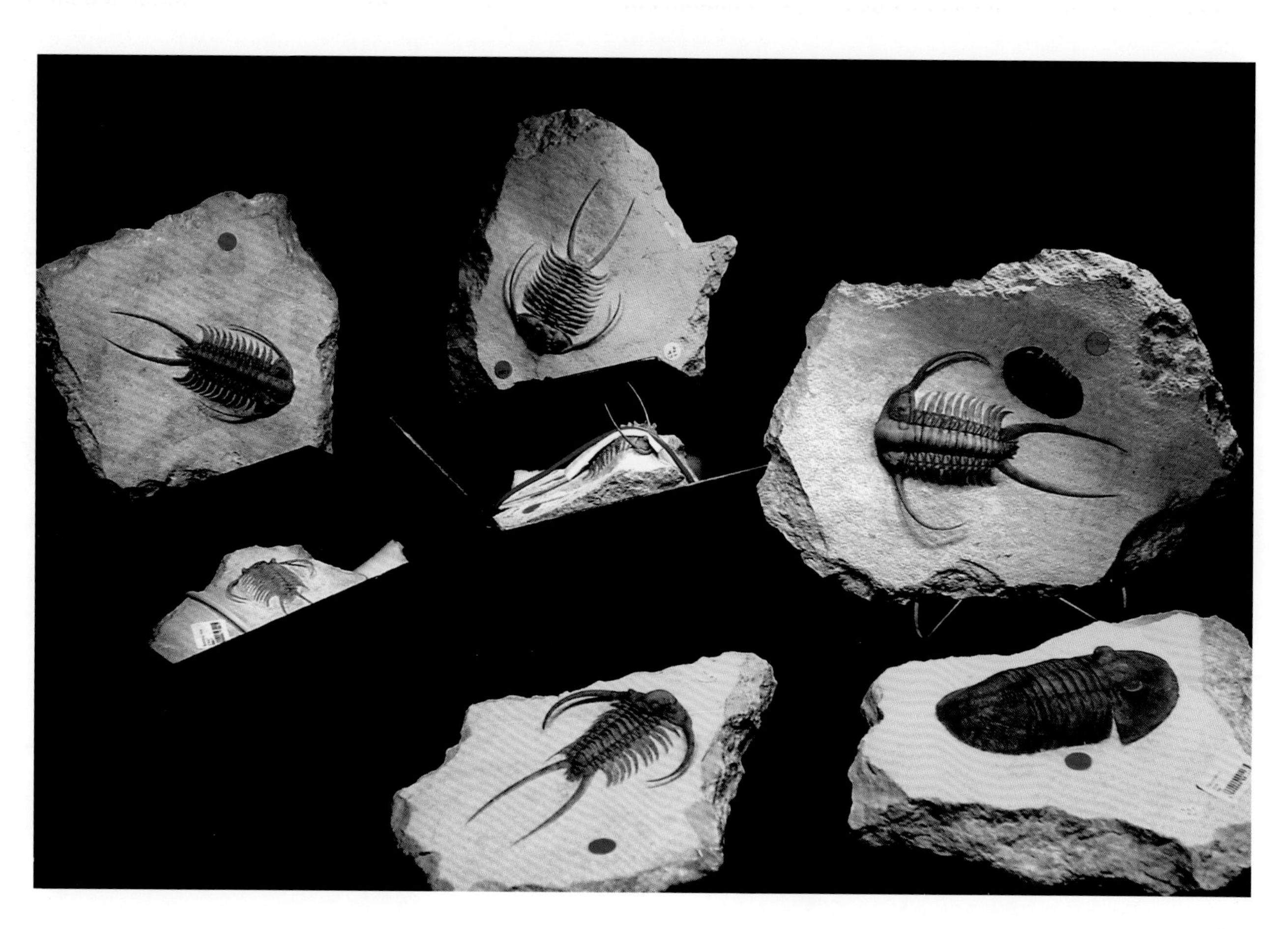

PLATE 216. A display of Russian Ordovician trilobites by the St. Petersburg Paleontological Laboratory, courtesy of Arkadiy Evdokimov, photographed by the author. Most of these are specimens of *Paraceraurus exsul* (Beyrich), in a variety of postures.

PLATE 218. This image contains three identical reproductions of a specimen of *Philonyx philonyx* Richter & Richter.

MOROCCAN TRILOBITE REPRODUCTIONS (FAKES)

PLATE 217. Unfortunately, the Tucson Show has become the outlet for a prospering industry of cheap but accurate reproductions, usually displayed in tents at the outskirts of the motels where the reputable dealers are exhibiting their precious wares. Here is an assembly of identical copies of *Psychopyge elegans* G. & H. Termier, to be sold at prices at least an order of magnitude lower than the genuine specimens seen at the show.

TRILOBITES FROM OTHER REPUTABLE WORLDWIDE SOURCES

PLATE 220. *Coronocephalus gauluoensis* Wu. (61 mm). Silurian, from Hu Bei, China. From Asian Link Trading at Ramada Ltd., Tucson. Specimen courtesy of Mike Hammer, photographed by the author.

PLATE 219. Box containing reproductions of more than one original of a variant (short trident) of *Walliserops trifurcatus* Morzadec.

PLATE 222. *Redlichia chinensis* Yung Shan (65 and 72 mm). L. Cambrian, Hunan Province, China. RLS coll. From Asian Link Trading at Ramada Ltd., Tucson.

PLATE 221. *Redlichia chinensis* Yung Shan (77 mm). L. Cambrian, Hunan Province, China. RLS coll. From Asian Link Trading at Ramada Ltd., Tucson. These trilobites belong to an important group, the order Redlichiida, progenitors of many radiating suborders and families of the Lower Cambrian. Among these, the basic structural and morphological characteristics of the earliest trilobites are known to have evolved.

PLATE 224. *Redlichia takooensis* Lu (98 and 79 mm). L. Cambrian, Emu Bay, Kangaroo Island, South Australia. Specimen courtesy of Norman Brown, photographed by the author. The long axial spine from the 11th axial ring, characteristic of this genus, is well evident in the longest specimen, while cut at the base in the specimen of the preceding plate.

PLATE 223. *Redlichia takooensis* Lu (59 mm). L. Cambrian, Emu Bay, Kangaroo Island, South Australia. RLS coll. From Ausrox at Inn Suites, Tucson.

3 THE EYES OF TRILOBITES

On March 3, 2010, I was pleasantly surprised to receive an e-mail message from my friend Nicholas (Nick) J. Pritzker containing a poem entitled "Lay of the Trilobite," written in 1887 by the Victorian poet Kay Kendall. Nick shares with me a keen interest in trilobites. Browsing the web on the topic of trilobite poetry, I also ran into a poem entitled "Ode to a Trilobite," written by another British poet, Timothy A. Conrad, in 1840. Both poets were captivated by the eyes of trilobites, and I deemed it pertinent to reproduce these poems in this context, in chronological order.

Ode to a Trilobite

Timothy A. Conrad (1840)

Thou large-eyed mummy of the ancient rocks,
The Niobe of ocean, couldst thou tell
Of thine own times, and of the earthquake shocks
Which tore the ocean-bed where thou didst dwell;
What dream of wild Romance would then compare
With the strange truths thy history might unfold?
How would Geologists confounded, stare
To find their glittering theories were not gold?

Methinks I see thee gazing from the stone
With those great eyes, and smiling as in scorn

Of notions and of systems which have grown
From relics of the time when thou wert born.
Which now in multiform profusion play,
Nor giant shells, nor monsters such as sweep
Along the surge and dash the ocean spray.

Yes, small in size were most created things
And shells and corallines the chief of these;
No land but islets then, nor trees nor springs,
And no tornado thundered o'er the seas.
But the wild earthquake did the work of death,
And heaped the sand and tore the Naiad's cave.
Race after race resigned their fleeting breath
The rocks alone their curious annals save.

And since the trilobites have passed away
The continent has been formed, the mountains grown
In oceans' deepened caves new beings play,
And Man now sits on Neptune's ancient throne.
The race of Man shall perish, but the eyes
Of Trilobites eternal be in stone,
And seem to stare about with wild surprise
At changes greater than they yet have known.

Lay of the Trilobite

Kay Kendall (1887)

A mountain's giddy height I sought,
Because I could not find
Sufficient vague and mighty thought
To fill my mighty mind;
And as I wandered ill at ease,
There chanced upon my sight
A native of Silurian seas,
An ancient Trilobite.

So calm, so peacefully he lay,
I watched him even with tears:
I thought of Monads far away
In the forgotten years.

How wonderful it seemed and right,
The providential plan,
That he should be a Trilobite,
And I should be a Man!

And then, quite natural and free
Out of his rocky bed,
That Trilobite he spoke to me
And this is what he said:
"I don't know how the thing was done,
Although I cannot doubt it;
But Huxley—he if anyone
Can tell you all about it;

"How all your faiths are ghosts and dreams,
How in the silent sea
Your ancestors were Monotremes —
Whatever these may be;
How you evolved your shining lights
Of wisdom and perfection
From Jelly-Fish and Trilobites
By Natural Selection.

"You've Kant to make your brains go round,
Hegel you have to clear them,
You've Mr Browning to confound,
And Mr Punch to cheer them!
The native of an alien land
You call a man and brother,
And greet with hymn-book in one hand
And pistol in the other!

"You've Politics to make you fight
As if you were possessed:
You've cannon and you've dynamite
To give the nations rest:
The side that makes the loudest din
Is surest to be right,
And oh, a pretty fix you're in!"
Remarked the Trilobite.

"But gentle, stupid, free from woe
I lived among my nation,

I didn't care—I didn't know
That I was a Crustacean.
I didn't grumble, didn't steal,
I never took to rhyme:
Salt water was my frugal meal,
And carbonate of lime."

Reluctantly I turned away,
No other word he said;
An ancient Trilobite, he lay
Within his rocky bed.
I did not answer him, for that
Would have annoyed my pride:
I merely bowed, and raised my hat,
But in my heart I cried: —

"I wish our brains were not so good,
I wish our skulls were thicker,
I wish that Evolution could
Have stopped a little quicker;
For oh, it was a happy plight,
Of liberty and ease,
To be a simple Trilobite
In the Silurian seas!"

As we now know, Kay Kendall's reference to trilobites as crustaceans, inhabiting primarily the Silurian seas, was replaced by better knowledge of modern taxonomy and geological age subdivisions. However, that knowledge was supported by the writings of Joachim Barrande, the grand master of paleontology of the time, who published in 1852 his magnificent monograph on trilobites referred to as *Crustacés,* part of his series entitled *Système Silurien du Centre de la Bohème.*

Like the above 19th-century poets, I was immediately captivated by the appearance of the compound eyes of trilobites. The fact that their eye lenses were made of calcite enabled their visual surfaces to be preserved with little change over hundreds of millions of years, without decay and consequent taphonomic alteration or destruction. Thus, we can still wonder at the spiraling, sunflower-seed-like arrangement of the eye lenses around the turret-like structures that encompass the lenses, as if a life-preserving guard against unknown perils. As introduced to me by the painstaking descriptions of their architecture by my colleague and friend Euan N. K. Clarkson, it was the internal structure of these lenses that aroused my physicist's attention, as briefly mentioned above in section 2.5 and in more detail in my previous trilobite books and joint publications. Thus, it became well established that the thick lenses in the aggregate eyes of a group of trilobites (Phacopidae) were doublet structures that eliminated spherical aberration in accordance with constructions by Descartes and Huygens (in their search for the best "burning glasses") (Clarkson and Levi-Setti 1975). A recent review paper entitled "The Eyes of Trilobites: The Oldest Preserved Visual System" (Clarkson, Levi-Setti, and Horvath 2006) summarizes various aspects of this topic. We were honored to be awarded a significant prize, the 50 Premio Internacional de Investigacion en Paleontologia (Paleonturologia '07) of the Fundacion Teruel of the Gobierno de Aragon. The paper was translated into Spanish and reprinted in an elegant, beautifully illustrated booklet.

In the context of this book, my journey in quest of trilobites brought me to exotic locations where I could actually unlock from the preserving rock a treasure of trilobite eyes to admire and illustrate. It would be repetitious for me to revisit the detailed description of the aspects and structures of the eyes of trilobites, since these are discussed in my earlier books. Furthermore, the advent and diffusion of web search engines makes immediately available to interested readers the most recent advances and literature on this topic as well.

PLATE 225. *Scabriscutellum (Cavetia) furciferum hamlaghdadianum* Alberti (eye length 11 mm). L. Devonian, Pragian, Hamar Laghdad Formation, Hamar Laghdad, near Erfoud, Morocco. RLS coll. Frontal view of a specimen I collected, devoid of any preparation. The corneal membrane of this holochroal eye is missing, revealing the fascinating compound eye architecture of the underlying visual surface. The closely packed ommatidia of each dorsoventral file are arranged along intersecting logarithmic spirals of opposite sense of coiling (see, e.g., Levi-Setti 1975, 1993). How this arrangement comes about, common in nature in the florets of the giant sunflower *Helianthus maximum* as well as in all compound eyes of trilobites and other arthropods, is mathematically explained in a book by Hermann Weyl entitled *Symmetry* (Weyl 1952).

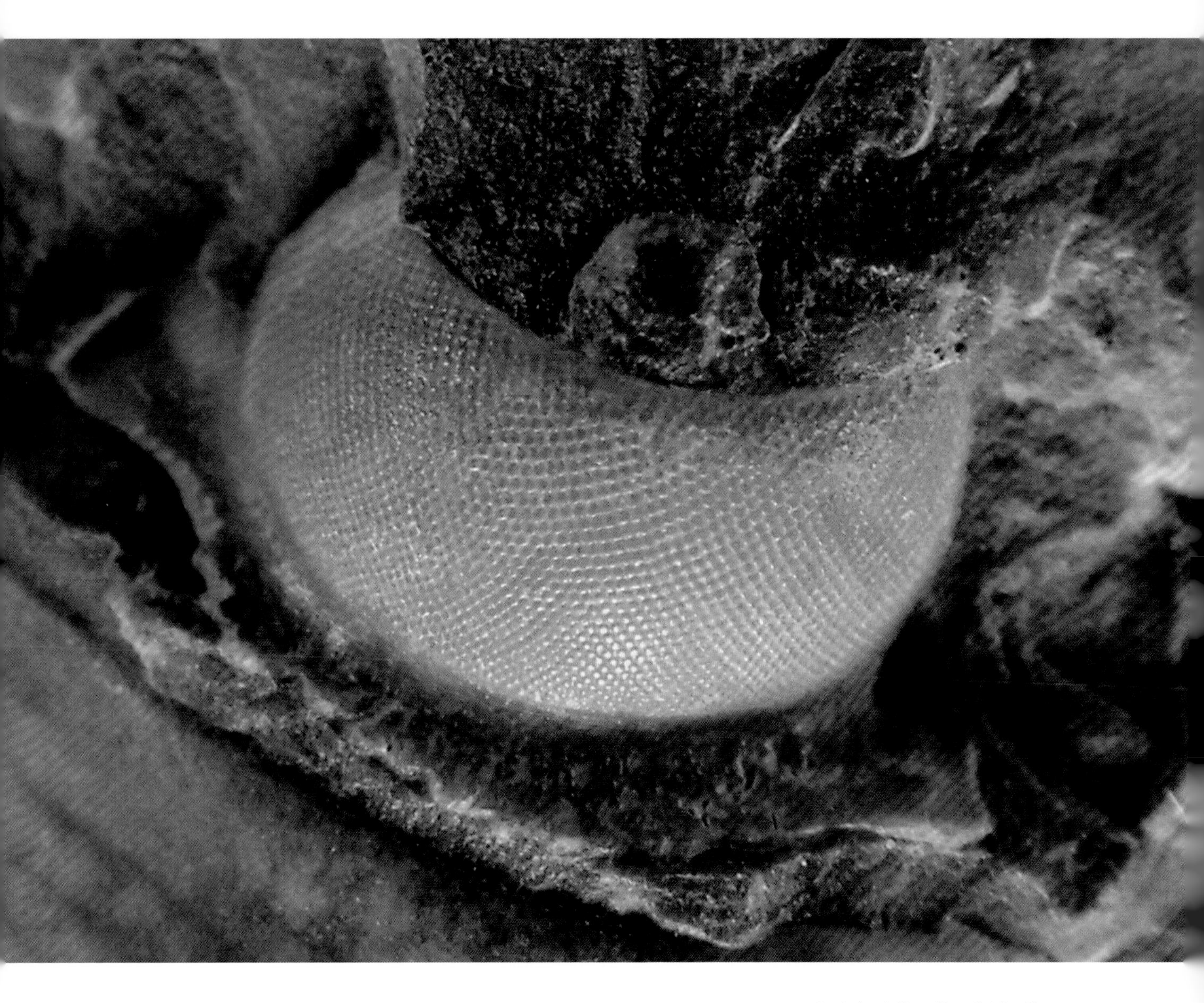

PLATE 226. *Scabriscutellum (Cavetia) furciferum hamlaghdadianum* Alberti (eye length 8 mm). L. Devonian, Pragian, Hamar Laghdad Formation, Hamar Laghdad, near Erfoud, Morocco. RLS coll. Angled view of another eye specimen of the same species as in the preceding plate, showing the toroidal shape of its entire visual surface. The bilateral positioning of such eye shapes clearly allowed a field of view covering the entire space surrounding the trilobite.

PLATE 227. *Phacops tafilaltensis* Alberti (eye length 10 mm). L. Devonian, Zguilma (southeast of Foum Zguid), Morocco. The schizochroal eye of this trilobite has only three or four vertically aligned ommatidia in each dorsoventral file, yet the intersecting logarithmic spiral arrangement of each unit can be detected. Each lens is surrounded by a clearly defined sclera.

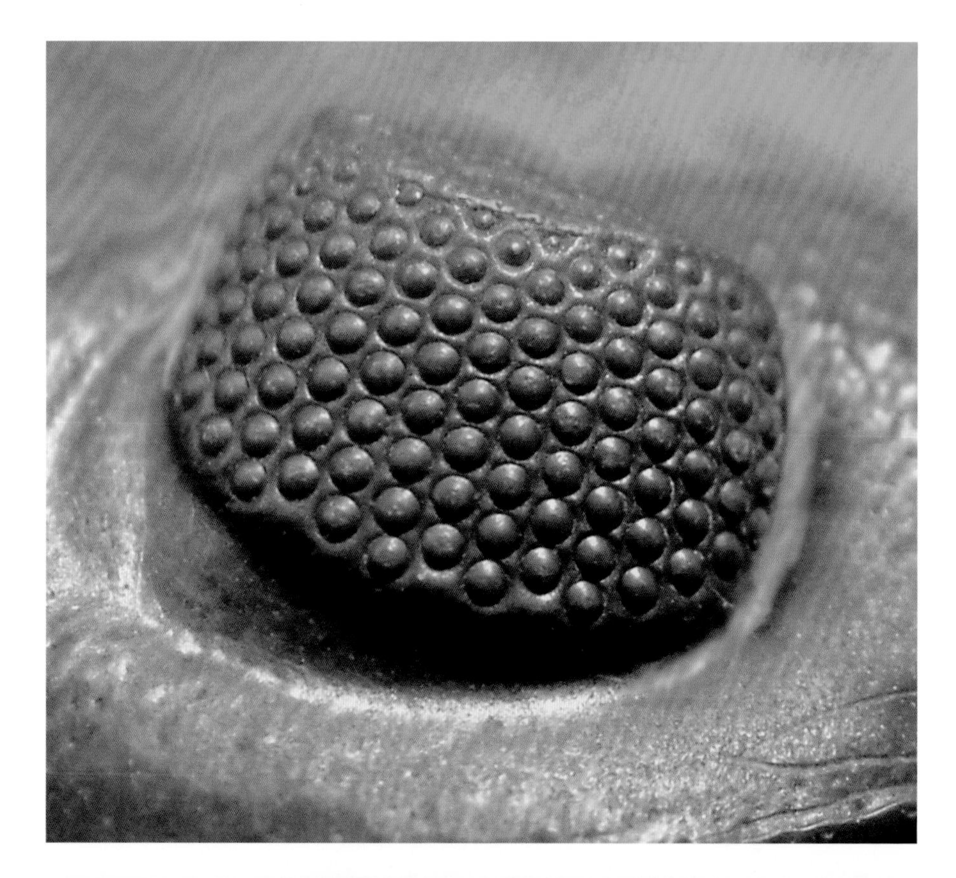

PLATE 228. *Eldredgeops rana milleri* Stewart, formerly *Phacops rana milleri* (eye length 9 mm). Devonian (Cazenovian), Silica Shale Formation, Hamilton Group, Medusa Portland Cement Quarry, Silica near Sylvania, Ohio. RLS coll. Beautifully preserved schizochroal eye of an enrolled specimen. In each dorsoventral file of seven units, the lens size is seen to gradually decrease from the bottom to the top of the eye where small lenslets appear. This reveals that the generative zone of the lenses resides at the top of the eye.

PLATE 229. *Eldredgeops rana crassituberculata* Stumm (eye length 9 mm). Devonian (Cazenovian), Silica Shale Formation, Hamilton Group, Medusa Portland Cement Quarry, Silica near Sylvania, Ohio. RLS coll. In this variant species, only five lenses make up each dorsoventral file, deeply embedded in swollen scleras.

PLATE 230. *Phacops (Drotops) megalomanicus* (Struve) (eye length 15 mm). M. Devonian, Jbel Issoumour, near Alnif, southeast Morocco. RLS coll. The architecture of this schizochroal eye is quite similar to that of *Eldredgeops rana crassituberculata* Stumm, although in enlarged aspect.

PLATE 231. *Hollardops mesocristata* (Le Maître) (eye length 9 mm). L. Devonian (Emsian), Timrarhart Formation, Jbel Anhsour, south of Foum Zguid, Morocco. RLS coll. A fine hexagonal decoration of dots is portrayed here on the sclera surrounding each lens. Similar decoration was first pointed out by Barrande (1852) in dalmanitid eyes. Even more evident than that noted in plate 228 is the progressive increase in width of the lenses in each dorsoventral file, away from the generative zone, following the increased lens spacing.

PLATE 232. *Odontochile (Zlichovaspis)* aff. *rugosa* Barrande (eye length 12 mm). M. Devonian, Jbel Issoumour, near Alnif, southeast Morocco. RLS coll. In this eye, many lenses have been dislodged, leaving empty alveoli.

PLATE 233. *Coltraneia oufatenensis* Morzadec (eye length 8 mm). L. Devonian (Emsian), El Otfal Formation, Oufatène, south of Jbel Issoumour, Ma'der, Morocco. RLS coll. In this fascinating eye, the calcite lenses, protruding well above the sclera, are translucent. Symmetrical light reflections from two side sources of illumination brighten the lenses of the central dorsoventral files. The latter contain 12 fully developed lenses each.

PLATE 234. *Erbenochile erbeni* (Alberti) (height of eye 8 mm). L. Devonian (Emsian), Timrarhart Formation, near Foum Zguid, Morocco. Photograph courtesy of Anna Librale of a specimen from the company Sahara Overland. The architecture of the visual surface of this trilobite differs significantly from that of any other schizochroal eye. The closely packed lenses, barely separated by a vanishing sclera (18 in each dorsoventral file), tower along a tall cylindrical surface terminating at its top to into a palpebral lobe that extends forward to form an eyeshade (Fortey and Chatterton 2003). Fortey and Chatterton have shown experimentally how effective this device can be in shielding the entire visual surface from parallel light impinging from above. This feature betrays a diurnal protective adaptation of the trilobite to detect preferentially motion occurring in the horizontal field of view, perhaps that of predators. One can speculate about a life habit where only the trilobite's turret eyes would be protruding from its hiding place buried under the cover of soft seafloor.

PLATE 235. *Erbenochile erbeni* (Alberti) (height of eye 8 mm). L. Devonian (Emsian), Timrarhart Formation, near Foum Zguid, Morocco. Specimen courtesy of Adam A. Aaronson, photographed by Pat Sword. This is a reenactment of the experiment shown by the authors mentioned in the preceding plate (Fortey and Chatterton 2003). The two views show the shadowing effect caused by the eyeshade upon slightly displacing the light source forward (in the lower view).

ACKNOWLEDGMENTS

My greatest thanks is owed to the late Susan Abrams, editor, University of Chicago Press, who first proposed the idea for the composition of this book. She was the one who suggested a more personal approach for me to share with other trilobite enthusiasts.

Christie Henry and Mark Reschke, current University of Chicago Press editors, were instrumental in helping me carry out Susan's concept. Thank you, Christie and Mark.

It is now 20 years since the second edition of my trilobite books, containing only black-and-white photographs, first appeared in print. The task of assembling a representative selection of digital color images of trilobites originating from the often-remote locations I visited in my field trips would have been a daunting enterprise without the contributions of many friends. Among these, my gratitude is extended to professional and well-known paleontologists, field guides, and companion diggers, amateur trilobite collectors, specimen preparators, and friendly dealers, familiar with, or met at such locations. Added to this list are close relatives and friends who have tolerated the hardship of accompanying me in my exploits as well as hosting me and my wife Nika in their residences for extended lengths of time.

Among professional paleontologists, I am much indebted to Prof. Dr. Gerd Geyer of the University of Würzburg, Germany, for providing me with detailed technical information about the geology and fauna of many locations of the Moroccan Anti-Atlas region. Over the years, he discussed with me the identification of my trilobite findings and kept me abreast of the latest developments in the field. Associated with him has been an indefatigable and knowledgeable amateur scientist,

Anthony Vincent, of Aberdeen, UK, who undertook a meticulous stratigraphic study of the Cambrian fauna present in the exposed layers among the quarries of the Tarroucht-Tarhia region of Morocco. It is in these quarries where I collected indiscriminately most *Acadoparadoxides* specimens in my collection, over many visits.

I remember fondly my association with Dr. Euan N. K. Clarkson of the Grant Institute of Geology of Edinburgh, Scotland, and the late Dr. Jan Bergström of the Swedish Museum of Natural History of Stockholm, Sweden, which resulted in several joint exploits mentioned in the acknowledgments of my previous trilobite books. More recently, Dr. Bergström also accompanied me on a memorable Morocco field trip.

Warm thanks are due to my expert field guide Abdullah Adam Aaronson who organized and made possible extended exploration of Moroccan fossil sites over many, sometimes hazardous lengthy tours (typically some 3,000 km long), through sandstorms and flash floods. His knowledge and expertise in trilobite digging, together with his command of Arabic dialects, have been an asset on many occasions. I am also indebted to another expert trilobite amateur and collector, Norman Brown of Sa Diego, California, who contributed to my image collection with photographs of rare complete Lower Cambrian specimens from the Latham Shale of the Marble Mountains of Southern California and the Middle Cambrian Chisholm Shale of the Half Moon Mine at Pioche, Nevada. He also guided me to explore the well-known fossil-rich exposures of the Lower Cambrian near Ploche, the Ruin Wash and Klondike Gap locality, and enabled me to photographs some of the precious specimens he collected there. Among close friends and colleagues who accompanied me and helped me search for trilobites in disparate locations, Mark Utlaut, now a physics professor at the University of Oregon, stands out for his enthusiasm and resourcefulness. Together, we went many times to the Marble Mountains in Southern California, the Chief Range in Nevada, to the House Range in Utah, in my early trips to Newfoundland, to Prague and the Bohemian Karst, and even on my first trio to Morocco.

A special relationship has developed between myself and Robert (Bob) Carroll, of Clarita, Oklahoma. He has credited me for inspiring him in his chosen profession upon discovering a copy of my first book in a library. Reciprocally, I now admire him for his craftsmanship in specimen preparation. He was instrumental in promoting my candidacy for the Charles H. Sternberg Award for Contributions to Paleontology by the Association of Applied Paleontological Sciences, which was granted to me at the 2010 Tucson Gem and Mineral Show. On a visit to his atelier in Clarita, he introduced me to the art of miniature air brazing delicate structures of spiny trilobites, his unmatched talent. His endeavors to publicize my present book are not to be forgotten. In my yearly visits at the Tucson Show, I also enjoyed my encounters with Bill Barker of the Sahara Sea Collection, who always allowed me to photograph the superb museum quality specimens in his exhibits, and for this warm thanks are due.

For many years, Judy and Richie Mease of Manuels, Newfoundland, have been instrumental in enabling me to carry out successfully my excavations of the giant *Paradoxides* trilobites in the Manuels River gorge. They offered not only hospitality in their beautiful home to me and my wife Nika, but Richie actually shared my digging efforts and offered his home workbench to me to facilitate my fossil preparation work. For all this essential assistance we are most grateful. As described in the introduction to the present book chapter devoted to Eastern Newfoundland, my activities in the Manuels River gorge came to the attention of the Manuels River Natural Heritage Society. One of the members of its Board of Directors, Pat Sword, became particularly active in promoting the construction of the now named Manuels River Hibernia Interpretation Centre, which now exhibits most of the trilobites I had collected from the banks of the Manuels River. As a professional photographer, Pat Sword documented not only my local excavations for illustration in the new center, but also accompanied me and Nika in my last field trip to Morocco and helped me in taking masterful photographs of trilobites encountered along our wanderings. Several of her pictures are shown in this book, and her many contributions are gratefully acknowledged here.

My excursions in search of trilobites have become family vacation time in most of my destinations, and my sons Emile and Matteo grew up helping me in my adventures. But my most heartfelt gratitude is felt for Nika, who provided me with endless assistance and support in often demanding circumstances. She endured the hardships of my extended travels into the Moroccan Anti-Atlas Pre-Sahara mountains and desert, and the many hours of waiting for my return from climbing explorations. Without protest, she let me fill every available space in our home with trilobites. It is due to her patience and encouragement that this book could be brought to completion.

REFERENCES

A. Specific Literature Referred to in the Text or Figure Captions

Alzenberg, J., A. Tkachenko, S. Weiner, L. Addadi, and G. Hendler. 2001. Calcitic microlenses as part of the photoreceptor system in brittlestars. *Nature* 412:819–822.

Barrande, J. 1852. *Système Silurien du Centre de la Bohême, Iere Partie: Recherces Paléontologiques*. Vol. 1. Praha-Paris.

Bergström, J., and R. Levi-Setti. 1978. Phenotypic variation in the Middle Cambrian trilobite *Paradoxides davidis* Salter at Manuels, SE Newfoundland. *Geologica et Palaeontologica* (Marburg) 12:1–40.

Bright, R. C. 1959. A paleoecological and biometric study of the Middle Cambrian trilobite *Elrathia kingii* (Meek). *J. Paleont.* 33:83–98.

Burkhard, H., and R. Bode. 2003. Trilobitenland Marokko. Keine angst vor fälschungen—*Offizielle Katalog der 40. Mineralientage München, Turmalin und Trilobit:* 138–144, München.

Clarkson, E. N. K., and R. Levi-Setti. 1975. Trilobite eyes and the optics of Descartes and Huygens. *Nature* 254:663–667.

Clarkson, E. N. K., R. Levi-Setti, and G. Horvath. 2006. The eyes of trilobites: The oldest preserved visual system. *Arthropod Structure & Development* 35:247–259.

Corbacho, J., and J. A. Vela. 2010. Giant trilobites from Lower Ordovician of Morocco. *Batalleria* 15:3–34.

Fortey, R. A. 2011. Trilobites of the genus *Dikelokephalina* from Ordovician Gondwana and Avalonia. *Geological Journal* 46(5):405–415.

Fortey, R. A., and Chatterton, B. 2003. A Devonian trilobite with an eye shade. *Science* 301:1689.

Geyer, G. 1993. The giant Cambrian trilobites of Morocco. *Beringeria* 8:71–107, Würzburg.

———. 1998. Intercontinental, trilobite-based correlation of the Moroccan early Middle Cambrian. *Can. J. Earth Sci.* 35:374–401.

Gould, S. J., and N. Eldredge. 1977. Punctuated equilibria: The tempo and mode of evolution reconsidered. *Paleobiology* 3:115–151.

Hansen, G. P. 2009. *Trilobites of Black Cat Mountain*. New York and Bloomington: iUnivers, Inc.

Hupé, P. 1953–1955. Classification des trilobites. *Ann. Paleontl.* 39(1953):59–168 and 41 (1955):91–325.

Levi-Setti, R. 1975. *Trilobites—A Photographic Atlas.* Chicago: University of Chicago Press.

———. 1993. *Trilobites.* 2nd ed. Chicago: University of Chicago Press.

Lieberman, B. S. 1999. *Systematic Revision of the Olenelloidea (Trilobita, Cambrian).* Bulletin 45 of the Peabody Museum of Natural History, Yale University, New Haven, CT.

Lu, Y.-H. 1957. Trilobita. In *Index Fossils of China*, ed. Nanjing Institute of Geology and Paleontology, Chinese Academy of Sciences, 249–294 (Picture Section, p. 155, figs. 11–13). Beijing: Geologica Publishing House.

Palmer, A. R. 1998. Terminal Early Cambrian extinction of the Olenellina: Documentation from the Pioche Formation, Nevada. *J. Paleont.* 72(4):650–672.

Palmer A. R., and Repina, L. N. 1993. *Through a Glass Darkly: Taxonomy, Phylogeny, and Biostratigraphy of the Olenellina.* The University of Kansas Paleontological Contributions (new series) 3:1–35. Lawrence: University of Kansas Paleontological Institute.

Šnajdr, M. 1958. The trilobites of the Middle Cambrian of Bohemia. *Rozpr. Ústřed. Ústavu Geol.* 24:1–280.

Vela, J. A., and J. Corbacho. 2009. Trilobites from the Upper Ordovician of Bou Nemrou-El Kaid Errami (Morocco). *Batalleria* 14:99–106.

Vincent, A. 2010. Private communication.

Webster, M. 2011. Trilobite biostratigraphy and sequence stratigrophy of the Upper Dyeran (traditional Laurentian "Lower Cambrian") in the Southern Great Basin U.S.A. In *Cambrian Stratigraphy and Paleontology of Northern Arizona and Southern Nevada,* ed. J. S. Hollingswort, F. A. Sundberg, and J. R. Foster. Museum of Northern Arizona Bulletin 67.

Webster M., R. R. Gaines, and N. C. Hughes. 2008. Microstratigraphy, trilobite biostratinomy and depositional environment of the "Lower Cambrian" Ruin Wash Lagerstätten, Pioche Formation, Nevada. *Palaeogeography, Palaeoclimatology, Palaeoecology* 264:100–122.

Weyl, H. 1952. *Symmetry.* Princeton, NJ: Princeton University Press.

B. Technical Literature Consulted in the Preparation of This Work

Berman, A., J. Hanson, L. Leiserowitz, T. F. Koetzle, S. Weiner, and L. Addadi. 1993. Biological control of crystal texture: A widespread strategy for adapting crystal properties to function. *Science* 259:776–779.

Brett, C. E. 2001. Fossils and fossilization. In *Encyclopedia of Life Sciences: 1–8.* New York: John Wiley & Sons, Ltd.

Briggs, D. E. G. 1999. Molecular taphonomy of animal and plant cuticles: Selective preservation and diagenesis. *Phil. Trans. R. Soc. Lond. B* 354:7–17.

Bruton, D. L. 2008. A systematic revision of *Selenopeltis* (Trilobita: Odontopleuridae) with description of new material from the Ordovician Anti Atlas region, Morocco. *Palaleontologische*

Zeitschrift 82(1):1–16.

Fletcher, T. P., G. Theokritoff, G. Stinson Lord, and G. Zeoli. 2005. The Early Paradoxidid Harlani trilobite fauna of Massachusetts and its correlatives in Newfoundland, Morocco, and Spain. *J. Paleont*. 79(2):312–336.

Gaines, R. R., M. J. Kennedy, and M. L. Droser. 2005. A new hypothesis for organic preservation of Burgess Shale taxa in the Middle Cambrian Wheeler Formation, House Range, Utah. *Palaeogeography, Palaeoclimatology, Palaeoecology* 229(1–2):193–205.

Gál, J., G. Horváth, and E. N. K. Clarkson. 2000. Reconstruction of the shape and optics of the lenses in the abathochroal-eyed trilobite *Neocobboldia chinclinica*. *Historical Biology* 14:193–204.

Gál, J., G. Horváth, E. N. K. Clarkson, and O. Himan. 2000. Image formation by bifocal lenses in a trilobite eye? *Vision Research* 40:843–853.

Geyer, G. 1996. The Moroccan fallotaspidid trilobites revisited. *Beringeria* 18:89–199. Würzburg.

Geyer, G., and A. R. Palmer. 1995. Neltneriidae and Holmiidae (Trilobita) from Morocco and the problem of the early Cambrian intercontinental correlation. *J. Paleont*. 69:459–474.

Geyer, G., and E. Landing. 2001. Middle Cambrian of Avalonian Massachusetts stratigraphy and correlation of the Braintree trilobites. *J. Paleont*. 75(1):116–135.

———. 2004. A unified Lower–Middle Cambrian chronostratigraphy for West Gonwana. *Acta Geologica Polonica* 54:179–218.

Gunther, L. F., and V. G. Gunther. 1981. Some Middle Cambrian fossils of Utah. *Brigham Young University Geology Studies* 28(part 1):1–87.

Howell, B. F. 1925. Faunas of the Cambrian *Paradoxides* beds at Manuels, Newfoundland. *Bull. Am. Paleont*. 43(11):1–140. Ithaca.

Hughes C. P., J. K. Ingham, and R. Addison. 1975. The morphology, classification and evolution of the trinucleidae (Trilobita). *Phil. Trans. R. Soc. Lond. B* 272:537–607.

Hunt, G., and R. E. Chapman. 2001. Evaluating hypotheses of instar-grouping in arthropods: A maximum likelihood approach. *Paleobiology* 27(3):466–484.

Hutchinson, R. D. 1962. Cambrian stratigraphy and trilobite faunas of southeastern Newfoundland. *Geol. Survey Canada* 88:1–156. Ottawa.

Jacobs, G. H., and J. Nathans. 2009. The evolution of primate color vision. *Sci. Am*. 4:56–63.

Jansen, U., G. Becker, G. Ploodowski, E. Schindler, O. Vogel, and K. Weddige. 2004. The Emsian and Eifelian near Foum Zguid (NE Draa Valley, Morocco). *Devonian of the Western Anti Atlas: Correlations and events. Doc. Inst. Sci, Rabat* 19:19–28.

Kim, D. H., S. R. Westrop, and E. Landing. 2002. Middle Cambrian (Acadian Series) conoryphid and paradoxididtrilobites from the Upper Chamberlain's Brook Formation, Newfoundland and New Brunswick. *J. Paleont*. 76(5):822–842.

Land, M. F. 1988. The optics of animal eyes. *Contemp. Phys*. 29(5):435–455.

Morzadec, P. 2001. Les trilobites asteropyginae du Dévonien de l'Anti–Atlas (Maroc). *Palaeontographica Abt. A* 262:53–84. Stuttgart.

Petrovich, R. 2001. Mechanisms of fossilization of the soft-bodied and lightly armored faunas of the Burgess Shale and of some other classical localities. *Am. J. Sci.* 301:683–726.

Pouget, E. M., P. H. H. Bomans, J. A. C. Goos, P. M. Frederik, G. de With, and Nico, A. J. M. Sommerdijk. 2009. The initial stages of template-controlled $CaCO_3$ formation revealed by cryo-TEM. *Science* 323:1455–1458.

Speyer, S. E. 1987. Comparative taphonomy and palaeoecology of trilobite lagerstätten. *Alcheringa: An Australasian Journal of Palaeontology* 11(3):205–232.

Struve, W. 1995. Die Reisen-Phacopiden auf den Maïder, SE-Marokkanische Prä-Sahara. *Senckerbergiana Lethaea* 75:77–129.

Thompson, K., K. May, and R. Stone. 1993. Chromostereopsis: A multicomponent depth effect? *Displays* 14(4):227–234. Butterworth-Heinemann Ltd.

Webster, M., P. M. Sadler, M. A. Kooser, and E. Fowler. 2003. Combining stratigraphic sections and museum collections to increase biostratigraphic resolution: Application to Lower Cambrian trilobites from southern California. In *Approaches in High Resolution Stratigraphic Paleontology,* ed. P. J. Harries, chap. 3. Dordrech, The Netherlands: Kluwer Academic Publishers.

Werner, J. S., B. Pinna, and L. Spillmann. 2007. Illusory color & the brain. *Sci. Am.* 3:91–95.

Wilmot, N. V., and A. E. Fallick. 1989. Original mineralogy of trilobite exoskeletons. *Palaeont.* 32(Part 2):297–304.

C. Web Sources and Books Consulted in the Preparation of This Work

Adobe Photoshop CS2 User Guide for Windows and Macintosh. 2005. San Jose, CA: Adobe Systems Inc. 416 pp.

Bonino, E., and C. Kier. 2010. *The Back to the Past Museum Guide to Trilobites.* Barzago (Lecco, Italy): Casa Editrice Marna s.c. 494 pp.

Chlupáč, I. 1993. *Geology of the Barrandian: A Field Trip Guide.* Senckerberg Buch 69. Frankfurt am Main: Senckerbergische Naturforschende Gesellschaft and the Czech State Geological Survey. 163 pp. ISBN 3-7829-1126-1.

Clarkson, E. N. K. 2008. *Invertebrate Palaeontology and Evolution.* 4th ed. Malden, MA, and Oxford: Blackwell Publishing. 452 pp. ISBN 978-0-632-05238-7.

Hardeberg, J. Y. 2001. *Acquisition and Reproduction of Color Images: Colorimetric and Multispectral Approaches.* San Antonio, TX: Universal-Publishers.com. 324 pp. ISBN 1-58112-135-0. See also related books on the web.

Gon, S. M., III. 2009. *The Trilobite Eye.* Web address: http://www.trilobites.info/eyes.htm,1–9.

Klikushin, V., A. Evdokimov, and A. Pilipyuk. 2009. *Ordovician Trilobites of the St. Petersburg Re-*

gion. St. Petersburg: Petersburg Paleontological Laboratory, Griffon Enterprises Inc., Master Fossil Japan. 539 pp.
Kowalski, H. 1992. *Trilobiten: Verwandungskünstler des Paläeozoikums*. Korb: Goldschneck-Verlag. ISBN 3-926129-12-3. 160 pp.
McEvoy, B. 2009. *Light and the Eye: Color Vision*. Web address: http://www.handprint.com/HP/WCL/color 1.html,1–69.
Moore, R. C., ed. 1959. *Treatise on Invertebrate Paleontology*. Part O. Lawrence: Geological Society of America and University Press of Kansas. 960 pp. This is the basic and most comprehensive reference on trilobites.
Moore, R. C., founder, and R. L. Kaesler, ed. 1997. *Treatise on Invertebrate Paleontology*. Part O, revised. Boulder, CO, and Lawrence: Geological Society of America and University Press of Kansas. 530 pp.
Šnajdr, M. 1990. *Bohemian Trilobites*. Prague: Geological Survey. 265 pp. ISBN 80-7075-001-4.
Weinmann, E., and P. Lourekas. 2001. *Photoshop 6 for Windows and MacIntosh*. Berkeley, CA: Peachpit Press. 516 pp. ISBN 0-201-71309-8.
Wikipedia, The Free Encyclopedia. 2009. *Trilobite*. Web address: http://en.wikipedia.org/wiki/Trilobite,1–17.
———. 2013. *Color Vision*. Web address: http://en.wikipedia.org/wiki/Color_¯vision, 1–13.

INDEX TO GENERA ILLUSTRATED IN THIS BOOK